MW00641341

PROFILES OF THE VACCINE INJURED

PROFILES OF THE VACCINE INJURED

"A LIFETIME PRICE TO PAY"

Children's Health Defense

Foreword by Robert F. Kennedy, Jr.

Skyhorse Publishing

Skyhorse Publishing books may be purchased in bulk at special discounts for sales promotion, corporate gifts, fund-raising, or educational purposes. Special editions can also be created to specifications. For details, contact the Special Sales Department, Skyhorse Publishing, 307 West 36th Street, 11th Floor, New York, NY 10018 or info@skyhorsepublishing.com.
Skyhorse® and Skyhorse Publishing® are registered trademarks of Skyhorse Publishing, Inc.®, a Delaware corporation.
Visit our website at www.skyhorsepublishing.com.
10 9 8 7 6 5 4 3 2 1

Library of Congress Cataloging-in-Publication Data is available on file.

Hardcover ISBN: 978-1-5107-7659-3
eBook ISBN: 978-1-5107-7660-9

Cover design by Brian Peterson
Printed in the United States of America

Contents

Acronyms

AAP	American Academy of Pediatrics
ABA	Applied Behavior Analysis
ACRP	Association of Clinical Research Professionals
ADHD	Attention-deficit/hyperactivity disorder
ALTE	Apparent life-threatening event
ARI	Autism Research Institute
ASD	Autism spectrum disorder
ASIA	Autoimmune/inflammatory syndrome induced by adjuvants
BMJ	Formerly, *British Medical Journal*
BRUE	Brief resolved unexplained event
CDC	U.S. Centers for Disease Control and Prevention
CFL	Compact fluorescent light bulb
CFS	Chronic fatigue syndrome
CHD	Children's Health Defense
CICP	Countermeasures Injury Compensation Program
CIDP	Chronic inflammatory demyelinating polyneuropathy
CNS	Central nervous system
D4CE	Doctors for Covid Ethics
DAN!	Defeat Autism Now!
DOJ	U.S. Department of Justice
DPT	Diphtheria-pertussis-tetanus (whole-cell pertussis)
DSM	*Diagnostic and Statistical Manual of Mental Disorders*
DT	Diphtheria-tetanus
DTaP	Diphtheria-tetanus-acellular pertussis
DTP	Diphtheria-tetanus-pertussis
DTPH	DTP+Hib
ENT	Ear, nose, and throat specialist

EPA	Environmental Protection Agency
ER	Emergency room
EUA	Emergency use authorization
FDA	U.S. Food and Drug Administration
FEMA	Federal Emergency Management Agency
FM	Fibromyalgia
FOIA	Freedom of Information Act
GAO	Government Accountability Office
GBS	Guillain-Barré syndrome
HHS	U.S. Department of Health and Human Services
Hib	*Haemophilus influenzae* type b
HMO	Health maintenance organization
HPV	Human papillomavirus
ICAN	Informed Consent Action Network
IEP	Individualized education program
IPV	Inactivated polio vaccine
IVIG	Intravenous immunoglobulin
LNP	Lipid nanoparticle
ME	Myalgic encephalomyelitis
MERS	Middle East respiratory syndrome
MMR	Measles-mumps-rubella
MMRV	MMR+varicella
MRI	Magnetic resonance imaging
mRNA	Messenger RNA
NAIP	National Adult Immunization Plan
NCVIA	National Childhood Vaccine Injury Act
NG	Nasogastric (feeding tube)
NIAID	National Institute of Allergy and Infectious Diseases
NIEHS	National Institute of Environmental Health Sciences
NIH	National Institutes of Health
NP	Nurse-practitioner
NVICP	National Vaccine Injury Compensation Program
OAP	Omnibus Autism Proceeding
OPV	Oral polio vaccine

OSHA	Occupational Safety and Health Administration
PA	Physician's assistant
PANDAS	Pediatric autoimmune neuropsychiatric disorder associated with streptococcal infections
PANS	Pediatric acute-onset neuropsychiatric syndrome
PCV	Pneumococcal conjugate vaccine
PDD	Pervasive developmental disorder
PDD-NOS	Pervasive developmental disorder-not otherwise specified
PEG	Polyethylene glycol
PI	Principal Investigator
PMD	Primary mitochondrial disease
PNS	Peripheral nervous system
POTS	Postural orthostatic tachycardia syndrome
PPO	Preferred provider organization
PREP Act	Public Readiness and Emergency Preparedness Act
SLE	Systemic lupus erythematosus
TB	Tuberculosis
VAERS	Vaccine Adverse Event Reporting System
VIS	Vaccine Information Statement
WHO	World Health Organization

Foreword

Vaccine injuries remain invisible to the majority of the public and sometimes even to the injured themselves. Government regulators and their industry captors have hidden their atrocities with the help of a craven fourth estate that long ago abdicated its watchdog role. Journalists, particularly science reporters, have abandoned journalism for stenography and a vicious form of propaganda that gaslights, marginalizes, and vilifies the injured and the honest doctors and scientists who report those injuries. They employ every alchemy of misdirection and deceit to silence dissent. Read the nine stories in this book, mirroring millions of similar injuries and deaths around the world. When you are done weeping and tearing out your hair from fury, frustration, and indignation, join Children's Health Defense in doing something about it.

—Robert F. Kennedy, Jr.

Prologue

MAKING VACCINE INJURY VISIBLE

In a world enamored with silver bullets and quick fixes, medical and public health officials, buttressed by decades of propaganda, have mesmerized much of the public into believing that vaccination reigns victorious and has no downside. Taking doctors' word as "gospel"[1] and deprived of meaningful informed consent,[2] few people pay even fleeting attention to vaccination risks—including health risks as well as the potentially devastating impact of an injury on a household's finances—when they bare an arm (their own or their child's) for one or more shots.

The controlled messaging,[3] together with censorship of vaccine injury stories in the public square[4] and "monstrous" gaslighting of injured individuals who persist in speaking up,[5] have thrown a cloak of invisibility over vaccination's potential to ruin health and torpedo financial security. Even in the face of soaring COVID-vaccine-related injuries and deaths,[6] these control tactics have ensured the ongoing relegation of vaccine adverse events to the shadows, leading the public to vastly underestimate risks.[7] A woman interviewed for this book commented that she had never stopped to reflect on how often vaccine injuries occur—but if she had, she would have assumed it was "even less than one in a million."

This book seeks to throw off the engineered invisibility cloak, shining an honest and unflinching light on vaccine damage and its effects on the lives not just of the injured but also their families and communities. The information we present underscores the following:

- Vaccine injuries are common, not rare.
- Vaccine injuries are "equal opportunity," affecting all demographic groups, including young and old, rich and poor.
- Vaccine injuries, more often than not, are profoundly life-changing.
- Experimental COVID injections have ramped up vaccine injuries to a level never seen before.

OVERVIEW OF CHAPTERS

In **Chapter One**, we summarize what we know—and still do not know—about vaccine injury. The fact that so many critical questions remain unanswered stems, in part, from suppression of the kind of research that would provide answers, as well as from intentional obfuscation by manufacturers and regulators of the unfavorable findings and data that do exist.

In **Chapter Two**, we discuss the limitations of vaccine injury compensation mechanisms and the often disastrous impact of an injury on individual and family finances, including both immediate family members and relatives such as grandparents.

In **Chapters Three through Six**, we present true stories of nine individuals injured by vaccines at ages ranging from infancy to middle age, situating their experiences in the broader context of the U.S. vaccination program and agenda. Adopting a wide-angle lens, each story communicates health impacts, the financial fallout, social impacts, and more. Sadly, the injuries described herein—and their painfully reverberating aftermath—are neither exaggerated nor exceptional. The stories include four children and adolescents harmed by vaccines on the Centers for Disease Control and Prevention (CDC) childhood schedule, two young people harmed (and in one case, killed) by COVID vaccines, and three adults injured by travel vaccines or by COVID shots.

In **Chapter Seven**, we take stock of the true risks of vaccination, discussing key themes and lessons learned from the nine stories and the scientific literature.

RIPPLE EFFECTS

When an individual becomes vaccine-injured, the effects, both short-term and long-term, can ripple out into all aspects of the person's and the family's life—with impacts on physical and mental health, finances, employment, marital and other family relationships, quality of life, longevity, and more. A vaccine injury becomes the perverse "gift that keeps on giving." To capture the various dimensions of this experience, we have organized each of our nine stories into the following sections:

- **Overview:** Here, we provide a summary of what happened.
- **Warning Signs:** In hindsight, the injured party or their family can often look back and spot warning signs or red flags that they themselves as well as health care workers failed to recognize, overlooked, or dismissed.
- **The Tipping Point:** Vaccine injuries are sometimes the result of cumulative vaccine exposures that build up (slowly or rapidly) over time until the individual reaches a tipping point. Whether the decline is gradual or abrupt, it usually can be linked to a particular shot or shots that acted as "the straw that broke the camel's back."
- **The Diagnosis:** Officialdom virtually never directly acknowledges damage caused by vaccines. Instead, vaccine-injured individuals generally have to run the gauntlet to obtain one or more conventional medical diagnoses that will allow them to access whatever support and resources may be available (such as health insurance coverage or Medicaid services).
- **Medical Experimentation:** Vaccines have always been an experiment, with many aspects—ingredients, combinations, synergistic effects, and much more—either never tested for safety at all, or tested with methods designed not to find anything. Our nine stories illustrate numerous aspects of this vast, unchecked experiment.
- **Day-to-Day Health Impacts:** Few people who have not experienced or witnessed a serious vaccine injury can fathom the day-to-day impact on health. Our stories provide a glimpse into the many challenges.

- **Financial Impacts:** The financial impact of a vaccine injury is one of the most invisible aspects of the injury experience. If parents, grandparents, and other relatives were more aware of the risk of bankruptcy that comes with a vaccine injury—and fully grasped the fact that manufacturers are immune from liability and that government compensation is far more unlikely than likely—people might weigh their vaccination decisions more carefully.
- **Social Impacts:** For a variety of reasons and in a variety of ways, vaccine injuries can be isolating. Shaking up a person's and a family's social life and social networks, injuries often create divisions between those willing to look vaccine damage squarely in the eye, and those who prefer to remain in denial about vaccination's risks.
- **Life Today:** Inevitably, a serious vaccine injury changes the course of someone's life.

MEET THE VACCINE-INJURED

In this book, you will meet the following:

TEMPLE—injured by childhood vaccines at 12 months of age.
Now a 22-year-old with severe autism, Temple's adverse reaction to vaccines came to a head when he was given a double dose of measles-mumps-rubella (MMR) vaccine concurrently with two other vaccines. At the time of Temple's vaccine injury, autism was so rare in the black community that Temple's mother had scarcely heard of it and knew of no one affected by it.

KEVIN—injured by prenatal biologics and childhood vaccines when 15 months old.
Now 27 years old, Kevin deteriorated at 15 months following repeated exposure to prenatal and postnatal vaccines and other biologics that contained hundreds of micrograms of the mercury-containing preservative thimerosal. At age three, Kevin was diagnosed with pervasive

developmental disorder (a subtype of autism). At age 21, he additionally began experiencing intractable seizures.

JACKSON—injured by childhood vaccines at 16 months of age.
Jackson, now 33, experienced a "swift and cruel" descent into pervasive developmental disorder and severe autism following vaccines administered when he was 16 months old. Between two and 16 months of age, Jackson received vaccines containing a total amount of mercury that was 139 times the Environmental Protection Agency (EPA) exposure level for an adult.

GIULIANA—injured by childhood vaccines through three years of age.
Giuliana, now a thriving 10-year-old, stopped breathing and turned blue right after receiving hepatitis B and vitamin K shots on the day of her birth; nurses told her parents this was "normal." When, at three months of age, she started having lapses of consciousness and turning bright red, doctors again dismissed it. An MMR shot at age three that prompted debilitating 48-hour migraines was the last straw, helping put Giuliana and the entire family on a different and healthier path.

MADDIE—injured in Pfizer's COVID vaccine clinical trial at age 12.
Before her injury as a participant in Pfizer's COVID vaccine clinical trial for 12- to 15-year-olds, Maddie was a motivated straight-A student and the strong and healthy "kid to watch" on her soccer team. Hours after receiving a second dose of Pfizer's experimental shot, Maddie began a downhill slide that has left her in a wheelchair with the muscle control of an infant, unable to take in nutrition other than through a feeding tube.

ERNESTO "JUNIOR"—killed by Pfizer's COVID vaccine at age 16.
A star athlete and only child, Ernesto "Junior" died in April 2021, five days after receiving one dose of Pfizer's COVID vaccine. His heartbroken single-parent father, who now "comes home to an empty house," is

"fighting for Junior's honor" by telling his son's story as widely as possible so that people understand the immense risks.

LISA—injured by travel vaccines at age 24.

A 48-year-old opera singer, Lisa experienced "cataclysmic" damage from travel vaccines in her mid-20s. Although years of detoxification and healing have allowed Lisa to regain most of her health, the injuries had numerous impacts on her operatic calling and career. Some symptoms persist 24 years later, and Lisa's doctors agree that she should never get another vaccine. Since COVID, however, no singer—no matter their medical condition—can "get through the door to audition without a vaccine card."

MONA—injured by Pfizer's COVID vaccine at age 41.

Mona was injured in April 2021 by a single dose of Pfizer's COVID vaccine. Before her vaccine injury, which put her in a wheelchair, Mona was a healthy mother of two who enjoyed going on bike rides with her 11-year-old daughter. The injury prevents Mona, who was in the middle of a career transition, from working and has turned her life "upside-down." She says, "I don't have a life right now because of what happened."

SUZANNA—injured by Pfizer's COVID vaccine at age 49.

A mother of two teenagers, Suzanna's COVID vaccine injury occurred in April 2021 following her second Pfizer shot. Before the vaccine injury, Suzanna was the family's higher-earning breadwinner and was extremely athletic—"healthier than your average 49-year-old." Two COVID shots have left her disabled, unable to work, confined to a wheelchair or walker, and in chronic pain. She says, "the majority of my waking moments are impacted by the injury."

CHAPTER ONE

What We Know (and Don't Know) about Vaccine Injury

AN UNHEALTHY NATION

By almost any measure, Americans are less healthy than their peers in other high-income nations. This health disadvantage, which "begins at birth and extends across the life course,"[8] has translated, especially in recent years, into plummeting life expectancy.[9]

Six in ten adults in the U.S. live with one or more chronic conditions,[10] and the situation is not appreciably different for children. Whereas in 1960, less than 2% of American children had health conditions "severe enough to interfere with usual daily activities,"[11] by 2007, an estimated 54% had at least one chronic health problem.[12] Some pediatricians believe that number is now far higher.

The start of the dramatic turn for the worse in American children's health can be traced to the late 1980s and early 1990s. It was also in the 1990s that researchers began issuing warnings about adults' worsening health trends and declining longevity. By the year 2000, researchers had pronounced the U.S. population as a whole a "curious outlier among its peers."[13]

VACCINES: THE ELEPHANT IN THE ROOM

To account for these dismal trends among both adults and children, researchers have shown themselves willing to explore some contributing factors but not others. Factors admitted for discussion include pervasive chemical exposures,[14] iatrogenic causes such as rampant opioid prescribing,[15] deteriorating food quality,[16] income inequality,[17] medical debt,[18] and lifestyle factors such as screen time.[19] However, the unmentioned elephant in the room—a standout culprit for children,[20] in particular—is vaccination. Mainstream medicine and media have managed to almost entirely embargo vaccination from discussion or consideration.

The timing of the downturn in children's health coincides uncannily with federal legislation and policy changes implemented in the late 1980s that triggered a substantial increase in the types and total number of vaccines required for school attendance. In the early 1980s, children received two dozen vaccine doses in their first 18 years, an already significant load; however, between 1989 and 2019, the number of total doses administered through age 18—not including prenatal shots—surged to roughly six dozen.[21] Stated another way, "a baby today receives more vaccinations by six months than her mother did by the time she graduated high school."[22]

Adults, too, have been pressured to accept a growing number of vaccines (see "Vaccination Cradle to Grave"), with mandates for COVID shots representing the latest widening of the dragnet.

VACCINATION CRADLE TO GRAVE

Prior to COVID, vaccination rates for children eclipsed uptake in American adults. However, public health officials have been champing at the bit to increase adult coverage. The *National Adult Immunization Plan* (NAIP) released in 2016, for example, included 78 strategies to catapult adult vaccination rates upward. In November 2020, building on the NAIP, the government drafted a *Vaccines National Strategic Plan* for 2021–2025, which renewed expressions of concern about "persistently low [adult] vaccination coverage rates."

If vaccination is contributing to the chronic disease epidemic in children, as a vast body of research suggests, it stands to reason that it is likely worsening adult health. Adults also face some unique vaccination risks. For example, adults receiving influenza, shingles, hepatitis B, and other vaccines have been the test group for the rollout of problematic new "smart" vaccine adjuvants, designed to ensure that even the most "mediocre" vaccine sends recipients' immune systems into overdrive. Moreover, in cohorts of younger adults, vaccines are piling on top of those adults' already heavy cumulative childhood vaccine exposures.

As of 2018, CDC reported that one in five U.S. adults had received every single vaccine (including flu shots) that CDC deems "age-appropriate." For individual vaccines, roughly half or more of age-eligible adults had received influenza (46%), tetanus (63%), and pneumococcal (69%) vaccines, and 53% of females aged 19-26 had received human papillomavirus (HPV) shots. The CDC claims that 86% of U.S. adults have received at least one COVID shot.

Sources:

Children's Health Defense. New adjuvants in the pipeline = more profits, questionable safety. Mar. 10, 2020. https://childrenshealthdefense.org/news/new-adjuvants-more-profits-questionable-safety/

Children's Health Defense. Vaxxed-Unvaxxed: The Science. https://childrenshealthdefense.org/wp-content/uploads/Vaxxed-Unvaxxed-Parts-I-XII.pdf

Glanz JM, Newcomer SR, Daley MF, et al. Cumulative and episodic vaccine aluminum exposure in a population-based cohort of young children. *Vaccine*. 2015;33(48):6736-6744.

Lu PJ, Hung MC, Srivastav A, et al. Surveillance of vaccination coverage among adult populations—United States, 2018. *MMWR Surveill Summ*. 2021;70(3):1-26.

Vaccines National Strategic Plan: 2021–2025. Department of Health and Human Services, Draft Nov. 13, 2020. https://www.hhs.gov/sites/default/files/Vaccines-National-Strategic-Plan-for-public-comment-2020-11-13.pdf

CHILDHOOD VACCINATION AND CHRONIC ILLNESS

American children are beleaguered by a long list of chronic afflictions[23]—sometimes nearly from birth. Publication in 2011 of a widely cited 2007 children's health survey showed that at that time, more than half (54%)

of American children had at least one of the following chronic health challenges:

- **Atopy**: asthma,[24] food/digestive allergies,[25] environmental allergies
- **Autoimmune**: diabetes[26]
- **Brain-related**: brain injury or concussion, epilepsy or seizure disorder, migraine headaches[27]
- **Developmental**: autism spectrum disorder (ASD),[28] developmental delays affecting ability to learn (or risk of developmental delay), learning disabilities, speech problems, Tourette syndrome
- **Eyes and ears**: chronic ear infections,[29] hearing problems, vision problems
- **Mental health**: anxiety problems, attention-deficit/hyperactivity disorder (ADHD),[30] behavior/conduct problems, depression[31]
- **Musculoskeletal**: bone, joint, or muscle problems[32]
- **Weight**: overweight/obesity[33]

Research spanning several decades strongly suggests that childhood vaccines have played a major role in constructing this picture of ill health. In fact, studies show that vaccinated children's health status is dramatically worse than that of similar groups of children not receiving the same vaccine(s),[34] and that vaccinated children face anywhere from a *two-fold to thirty-fold* increased risk of developing various acute and chronic conditions (see Table 1, which compiles studies on selected health outcomes).

Children's Health Defense (CHD) has been unable to find any studies showing that vaccinated children have superior health outcomes compared to their unvaccinated peers.[35]

TABLE 1. INCREASED RISK OF ADVERSE HEALTH OUTCOMES IN VACCINATED (VERSUS UNVACCINATED) CHILDREN, BY TYPE OF VACCINE (WITH CITATIONS)	
Acute respiratory infections	Childhood vaccines (Lyons-Weiler & Thomas, 2020[36]) Influenza vaccines (Rikin et al., 2018[37])

Aggressive behavior	Childhood vaccines (NVKP Survey, 2004[38])
Allergic reactions, allergic rhinitis, allergies	Childhood vaccines (Mawson et al., 2017a;[39] NVKP Survey, 2004) DTP and tetanus (Hurwitz & Morgenstern, 2000[40])
Anemia	Childhood vaccines (Lyons-Weiler & Thomas, 2020)
Asthma, asthma-related hospitalization	Childhood vaccines (Enriquez et al., 2008;[41] Lyons-Weiler & Thomas, 2020; NVKP Survey, 2004) Vaccines before age 1 (Hooker & Miller, 2020[42]) Inactivated vaccines (Yamamoto-Hanada et al., 2020[43]) DPT (McDonald et al., 2008[44]) HPV (Geier et al., 2019[45]) Influenza (Joshi et al., 2009[46])
Atopy	Measles (Shaheen et al., 1996[47])
ADHD	Childhood vaccines (Mawson et al., 2017a) Hepatitis B (Geier et al., 2018[48])
Autism	Childhood vaccines (Mawson et al., 2017a) DTaP (Geier et al., 2013[49]) Hepatitis B (Gallagher & Goodman, 2010;[50] Geier et al., 2013; Verstraeten Gen. 1, 2000[51]) MMR (DeStefano et al., 2004, unpublished data)[52]
Behavioral issues	Childhood vaccines (Lyons-Weiler & Thomas, 2020)
Bell's palsy	H1N1 influenza (Bardage et al., 2011[53])
Cardiac events (preemies)	Single/multiple vaccines (Pourcyrous et al., 2007[54])
Celiac disease	HPV (Hviid et al., 2018[55])
Chorioamnionitis	Tdap (Kharbanda et al., 2014[56])
Chronic lung disease	Childhood vaccines (NVKP Survey, 2004)
Conjunctivitis	Childhood vaccines (Lyons-Weiler & Thomas, 2020)
Convulsions, febrile convulsions, collapse	Childhood vaccines (NVKP Survey, 2004)
Crohn's disease	Measles (Thompson et al., 1995[57]) Polio (Pineton de Chambrun et al., 2015[58])

Developmental delays	Vaccines before age 1 (Hooker & Miller, 2020)
Diabetes, type 1	Pediatric vaccines (Classen, 2008[59]) BCG (Classen & Classen, 2003[60]) Hepatitis B (Classen & Classen, 1997[61]) MMR (Classen & Classen, 2003) Pertussis (Classen & Classen, 2003)
Ear infections	Childhood vaccines (Lyons-Weiler & Thomas, 2020; Mawson et al., 2017a; NVKP Survey, 2004) Vaccines before age 1 (Hooker & Miller, 2020)
Eating disorders	Childhood vaccines (Lyons-Weiler & Thomas, 2020)
Eczema	Childhood vaccines (Lyons-Weiler & Thomas, 2020; Mawson et al., 2017a; NVKP Survey, 2004) Inactivated vaccines (Yamamoto-Hanada et al., 2020)
Emotional disturbances	Hepatitis B (Geier et al., 2017[62])
Fetal loss	H1N1+seasonal influenza (Goldman, 2013[63])
Fever	Childhood vaccines (Lyons-Weiler & Thomas, 2020)
Gastrointestinal disorders/ gastroenteritis	Childhood vaccines (Lyons-Weiler & Thomas, 2020) Vaccines before age 1 (Hooker & Miller, 2020)
Infant mortality	Vaccines in infancy (Kristensen et al., 2000;[64] Miller & Goldman, 2011[65]) DTP (Aaby et al., 2004;[66] Mogensen et al., 2017[67]) Early DTP (girls) (Aaby et al., 2012;[68] Aaby et al., 2018[69]) DTP+BCG (infant girls) (Moulton et al., 2005[70]) DTP+measles (Aaby et al., 2015[71])
Inflammatory bowel disease	H1N1 influenza (Bardage et al., 2011)
Intussusception	Rotavirus (Patel et al., 2011[72])
Involuntary movement	HPV (Yaju & Tsubaki, 2019[73])
Learning disability	Childhood vaccines (Mawson et al., 2017a)
Liver problems	Hepatitis B (Fisher & Eklund, 1999[74])
Memory impairment	HPV (Yaju & Tsubaki, 2019)

Multiple sclerosis	Hepatitis B (Hernán et al., 2004[75])
Narcolepsy	H1N1 influenza (Miller et al., 2013;[76] Szakács et al., 2013[77])
Neurodevelopmental disorders (NDDs), NDDs in children born preterm	Childhood vaccines (Mawson et al., 2017a) ≥ 1 recommended vaccine (Mawson et al., 2017b[78]) Hepatitis B (Verstraeten Gen. 1, 2000)
Noninfluenza respiratory infection	Influenza (inactivated) (Cowling et al., 2012[79]) Influenza (Dierig et al., 2014;[80] Kelly et al., 2011;[81] Khurana et al., 2013[82])
Paraesthesia	H1N1 influenza (Bardage et al., 2011)
Pneumonia	Childhood vaccines (Mawson et al., 2017a)
Premature puberty	Thimerosal-containing vaccines (Geier et al., 2010[83]) Hepatitis B (Geier et al., 2018[84])
Sleep disorders	Childhood vaccines (NVKP Survey, 2004) Hepatitis B (Verstraeten Gen. 1, 2000)
Special education, receiving	Hepatitis B (Gallagher & Goodman, 2008[85])
Speech disorders	Hepatitis B (Verstraeten Gen. 1, 2000)
Sudden infant death syndrome	DPT (Torch, 1982[86])
Throat inflammation	Childhood vaccines (NVKP Survey, 2004)
Tics (boys)	Thimerosal-containing vaccines (Thompson et al., 2007[87])
Ulcerative colitis	Measles (Thompson et al., 1995) Polio (Pineton de Chambrun et al., 2015)

For further reading, see the Children's Health Defense e-book: *The Sickest Generation: The Facts Behind the Children's Health Crisis and Why It Needs to End*

A LIABILITY-FREE INDUSTRY

The catalyst for the precipitous expansion of the childhood and adolescent vaccine schedules that began three decades ago in the U.S. was the National Childhood Vaccine Injury Act (NCVIA). Passed into law in

1986, this devastating piece of legislation essentially abolished vaccine injury lawsuits against manufacturers and health providers, establishing blanket liability protections that removed any incentive for vaccine companies to make their products safe.[88] A 2011 Supreme Court decision further cemented the NCVIA's no-liability pledge.[89]

Catherine Austin Fitts, founder and president of Solari Inc. and a former senior government official and investment advisor, explains the ramifications of the liability protections as follows:

> *"Call a drug or biotech cocktail a 'vaccine,' and pharmaceutical and biotech companies are free from any liabilities—the taxpayer pays. Unfortunately, this system has become an open invitation to make billions from 'injectables,' particularly where government regulations and laws can be used to create a guaranteed market through mandates. As government agencies and legislators as well as the corporate media have developed various schemes to participate in the billions of profits, significant conflicts of interest have resulted."*[90]

In the U.S., emergency use authorization (EUA) medical "countermeasures"—including current mRNA and other COVID vaccines—enjoy their own set of sweeping liability protections[91] set up by the 2005 Public Readiness and Emergency Preparedness (PREP) Act. The PREP Act "immunizes" vaccine manufacturers and others from legal liability "for all claims for loss relating to the administration or use of a covered countermeasure."[92] "Losses" for which manufacturers and others face zero liability include death; physical, mental, or emotional injury, illness or disability; and "loss of or damage to property, including business interruption loss." Outside the U.S., manufacturers of COVID shots have obtained similar liability protections "through contractual and/or legislative efforts."[93]

The Food and Drug Administration's (FDA's) EUA authority was granted through legislation (the Project Bioshield Act of 2004) passed in the aftermath of 9/11 that allows FDA to authorize formally unapproved products for "emergency use" against a threat to public health and safety, subject to a declaration of emergency by the Department of Health and

Human Services (HHS).[94] Before COVID, FDA used this power "relatively sparingly," according to a report by Harvard Law.[95] In 2009, FDA authorized 22 non-vaccine EUA countermeasures for the H1N1 swine flu. In later years, it "preemptively" authorized a handful of experimental medical countermeasures for Middle East respiratory syndrome (MERS), Ebola, and Zika. Prior to COVID, however, FDA had granted EUA status to only one vaccine, a repurposed anthrax vaccine.

COVID marked a dramatic turning point in the FDA's wielding of the EUA mechanism—a veritable opening of the floodgates. Since March 2020, FDA has issued almost 400 COVID-related EUAs—"for personal protective equipment, medical equipment, in vitro diagnostic products, drug products, and, most notably, [brand-new] vaccines."[96]

For further reading, see the Children's Health Defense e-book:
Conflicts of Interest Undermine Children's Health

THE VACCINE ADVERSE EVENT REPORTING SYSTEM (VAERS)

Although the 1986 Act established a statutory obligation for HHS to improve vaccine safety and report on its progress biannually to Congress, HHS—in direct violation of Federal law—has never once done so in over 30 years.[97] As another commitment legislated through the NCVIA, the FDA and the CDC also pledged to monitor vaccine safety through the Vaccine Adverse Event Reporting System (VAERS), launched in 1990.[98] However, this voluntary reporting system's well-documented problems make it virtually impossible to get a meaningful handle on the extent and scale of vaccine injuries in the U.S.[99]

FDA and CDC are well aware of—and in fact play up—VAERS' shortcomings but have taken no steps to improve data collection and analysis. On the contrary, the two agencies seem more interested in letting VAERS remain "broken," languishing in obscurity and remaining unknown to many members of the public and even to the very health care providers[100] supposedly required by law to report vaccine adverse events.[101]

As a solitary exception to this institutional inertia, HHS commissioned a study (2007–2010) over a decade ago to assess the extent of

VAERS underreporting.[102] The Harvard Pilgrim Health Care consultants hired by HHS, engaging in a rigorous attempt to zero in on the unknowns, found that VAERS represented the tip of a very large injury iceberg, *capturing "fewer than 1%" of vaccine adverse events* (see "The Unreported 99 Percent").

THE UNREPORTED 99 PERCENT

In 2010, government consultants estimated that more than 99% of vaccine injuries are *not reported* to VAERS. Why? Researchers acknowledge numerous reasons, including the following:

- **Lack of awareness** of the voluntary reporting system
- Failure, by both health care providers and vaccine recipients, to **connect the dots** between vaccination and subsequent adverse impacts
- Wide variability in what is considered worthy of reporting, depending on **type of symptom** and the **timing** of symptom onset
- A cumbersome user interface, clinical time constraints, and other **logistic barriers**
- A **medical culture** that has never made adverse event reporting an "ingrained practice"
- Actual **institutional disincentives** for physicians to acknowledge or report vaccine injuries

Sources:

Baker MA, Kaelber DC, Bar-Shain DS, et al. Advanced clinical decision support for vaccine adverse event detection and reporting. *Clin Infect Dis.* 2015;61(6):864-870.

Kessler DA. Introducing MEDWatch: A new approach to reporting medication and device adverse effects and product problems. *JAMA.* 1993;269(21):2765-2768.

Rosenthal S, Chen R. The reporting sensitivities of two passive surveillance systems for vaccine adverse events. *Am J Public Health.* 1995;85(12):1706-1709.

The Highwire. "These patients deserve to be heard" – VAERS whistleblower. Sep. 17, 2021. https://thehighwire.com/videos/these-patients-deserve-to-be-heard-vaers-whistleblower/

ONE IN THIRTY-NINE

The Harvard consultants also made an effort to get a fix on the true prevalence of vaccine injury. After piloting a model surveillance system that analyzed data on hundreds of thousands of patients and over a million vaccine doses, they estimated that roughly 1 in every 39 vaccine doses administered—or 2.6%—produced an adverse event. When some of the researchers went on to test the pilot system in a large health network in Ohio, adverse event reporting increased by a factor of 30.[103] However, after learning of the alarming levels of vaccine injury exposed by the beefed-up system, CDC killed the initial study and thereafter refused to take phone calls from the consultants—even though the researchers had unearthed the very information CDC ostensibly had tasked them with obtaining.[104]

To this day, HHS and CDC remain mum about their crucial study, instead continuing to describe vaccine adverse events as "exceedingly rare."[105] The nation's lead immunologist, Dr. Anthony Fauci—long-time director of the National Institute of Allergy and Infectious Diseases (NIAID)—likewise routinely mischaracterizes vaccine adverse events as so infrequent as to be "almost nonmeasurable."[106]

Even with over 99% of vaccine adverse events likely going unreported, a daunting number of injury reports have been submitted to VAERS since 1990, putting the lie to HHS's and Fauci's disingenuous statements and showing that vaccine injury is both a long-standing and widespread phenomenon. Through November 2020—that is, until just prior to the COVID vaccine rollout—VAERS reports numbered nearly a million,[107] including 12,620 deaths.[108] Those numbers would be orders of magnitude higher if the agencies responsible for overseeing the flawed system were actually interested in capturing real data.

It is also important to note that adverse events do not necessarily occur in the immediate aftermath of vaccination. The potential for delayed reactions and slowly simmering vaccine-related illness—manifesting days,[109] weeks,[110] months,[111] or even years later[112]—makes it even more challenging to recognize adverse events and likely contributes to extensive underreporting.

INJURY CLUES IN CHILDHOOD VACCINE PACKAGE INSERTS

All vaccines come with manufacturer-authored package inserts that summarize product indications and usage, dosage, contraindications, warnings and precautions, adverse reactions, drug interactions, use in specific populations, clinical pharmacology, non-clinical toxicology, clinical trial findings, and more.[113] Though the inserts do not provide information about how often vaccine injuries occur, the data they compile from clinical trials and post-marketing reports demonstrate the occurrence of a wide range of injuries in both the short and longer terms.[114] Moreover, the post-marketing information included in the inserts is selective, not comprehensive; manufacturers get to choose what to include, and though they profess to include information based on "severity, frequency of reporting or strength of evidence for a [vaccine] causal relationship,"[115] there is no way to assess the level of potential cherry-picking.

The CDC's childhood and adolescent vaccine schedules include more than three dozen different vaccine brands made by eight manufacturers. The remarkable information squirreled away in the vaccines' package inserts discloses roughly 400 different types of documented adverse reactions.[116]

Taken as a whole, vaccine package inserts tell us the following:

- All vaccines are capable of causing—and do cause—adverse events.
- Some of the adverse events include the very illnesses (and related consequences) that the vaccines are supposed to prevent.
- Vaccine adverse events can affect any body system, including the all-important immune and nervous systems.
- The adverse reactions documented in clinical trials provide only a partial and very short-term picture of possible reactions.
- Many adverse events listed in package inserts are serious, potentially lifelong, or fatal.

For further reading, see the Children's Health Defense two-part series on vaccine package inserts:

- Read the fine print: vaccine package inserts reveal hundreds of medical conditions linked to vaccines (April 14, 2020)
- Read the fine print, part two—nearly 400 adverse reactions listed in vaccine package inserts (August 14, 2020)

WORLD HEALTH ORGANIZATION BEHIND-CLOSED-DOORS ADMISSIONS

In early December 2019, vaccine professionals from around the world attended a two-day World Health Organization (WHO) Global Vaccine Safety Summit. The comprehensive list of attendees included members of the Global Advisory Committee on Vaccine Safety (past and present), vaccine program managers, regulators, drug safety staff, academics, and representatives of United Nations agencies, donor agencies, and the pharmaceutical industry. In astonishing revelations caught on camera,[117] the world's top vaccine scientists admitted the following: vaccines can be fatal; vaccine adjuvants increase risk but are in most injections; vaccine safety science and monitoring is utterly inadequate and often "obfuscates" serious adverse events; and accordingly, confidence in vaccines is waning.[118]

One of the attendees was a program manager from Nigeria, Dr. Bassey Okposen, who ventured to ask whether vaccines containing "different antigens from different companies" and "different adjuvants and different preservatives and so on" might be "cross-reacting amongst themselves"—and wondered aloud whether "the possibility of cross-reactions" had ever been studied. In the U.S., the answer is a resounding "no." The Institute of Medicine (IOM) reported a decade ago that no research exists on "key elements of the entire [childhood vaccine] schedule—the number, frequency, timing, order, and age at administration of vaccines"[119]—and the intentional research gaps have only become more glaring since that time.

At the summit, behavioral scientist Heidi Larson, who has made a career out of designing strategies to overcome "vaccine hesitancy," acknowledged the widespread mistrust, including among "a very wobbly health professional front line that is starting to question vaccines and the safety of vaccines." Although Larson's wheelhouse is "contextual issues and communication issues," she also stated, "You can't repurpose the same old science to make it sound better if you don't have the science that's relevant to the new problems."

A handful of months later, the many unsettling questions bluntly raised by the WHO insiders had been almost entirely eclipsed by splashy announcements about a new generation of "Warp Speed" COVID vaccines, accompanied by a sizable helping of fear-mongering designed to obliterate any "vaccine hesitancy" people previously might have entertained.

NEW VACCINE TECHNOLOGIES, NEW RISKS

Since 2020's abbreviated clinical trials, the FDA has granted EUAs for four COVID-19 vaccines, and in June 2022, went so far as to extend two of the shots' authorized use down to infants as young as six months old.[120] FDA granted EUA status to:

1. The messenger RNA (mRNA) injection developed by Pfizer with German partner BioNTech—authorized for ages six months and up.
2. The mRNA injection developed by Moderna in partnership with NIAID—also authorized for ages six months and up.
3. The adenovirus-vectored shot made by Johnson & Johnson (J&J) subsidiary Janssen, authorized for age 18 and up but downgraded in May 2022 for use only in "certain individuals" due to a rare FDA admission that recipients risk life-threatening blood clots.[121] (FDA has not conceded the point that both the Pfizer and Moderna shots come with dangerous blood clot risks as well.[122])
4. The recombinant moth-cell-based Novavax vaccine, featuring a never-before-approved nanoparticulate adjuvant,[123] authorized for age 12 and up.[124]

Although FDA went on to formally approve the Pfizer injection (on August 23, 2021) under the brand name "Comirnaty" (for ages 16 and up) and the Moderna injection (on January 31, 2022) under the brand name "Spikevax" (for ages 18 and up), these "ghost vaccines" are not actually available for U.S. civilians,[125] meaning that most Americans continue to receive EUA injections, not licensed vaccines.[126] Whistleblowers report that in June 2022, vaccine vials labeled "Comirnaty" suddenly began appearing "unannounced" at military facilities—but they came from a manufacturing facility not approved by FDA.[127]

The first three EUA injections are gene therapy,[128] sharing the end goal of getting genetic instructions into a person's cells and "tricking" the cells into making coronavirus spike protein. At the World Health Summit in Berlin in October 2021, the head of Bayer's Pharmaceuticals Division, Stefan Oelrich, emphasized the importance—for industry—of the COVID mRNA vaccines, which established a crucial precedent for novel "cell and gene therapies" in the previously skeptical public eye. Oelrich delightedly stated, "If we had surveyed, two years ago, in the public, 'Would you be willing to take gene or cell therapy and inject it into your body?' we would have probably had a 95% refusal rate. I think this pandemic has. . . opened many people's eyes to innovation in the way that was not possible before."[129]

The lipid nanoparticles (LNPs) used in both mRNA shots as an adjuvant and "carrier" system to drive the synthetic mRNA into recipients' cells can cross the blood-brain barrier and accumulate in other vital organs;[130] the LNPs also are coated with a controversial polymer called polyethylene glycol (PEG) about which CHD has been issuing warnings since before the EUAs were granted.[131] According to the expert group Doctors for Covid Ethics (D4CE), the mRNA technology is so inherently dangerous that the COVID shots "should never even have been introduced."[132] With respect to the J&J shot, D4CE states that "introducing DNA into human cells" raises the possibility of "stable, irreversible incorporation into the human genome."[133]

The rollout of these gene therapy technologies has been nothing short of disastrous. A few months of clinical trial data (through February

28, 2021) submitted by Pfizer to FDA in support of COVID-19 vaccine licensing listed 42,086 adverse event case reports, with four-fifths (83%) originating in the U.S. The case reports involved close to 160,000 distinct adverse events, an average of nearly four per case.[134] Subsequent disclosures indicate that within three months of the shot's EUA launch,[135] Pfizer was hiring hundreds of additional employees "just to process the flood of adverse events reported."[136]

Adverse vaccine reactions reported to Pfizer included anaphylaxis, facial paralysis, autoimmune reactions, and pregnancy-related impacts, as well as adverse events involving the cardiovascular, dermatological, hematological, hepatic, musculoskeletal, neurological, renal, respiratory, thromboembolic, vasculitic, and "other" body systems (see "Primary 'System Organ Classes' Affected by Pfizer Shots"). An appendix supplied by Pfizer enumerated 1,291 possible "adverse events of special interest" following its COVID shot.[137] However, Pfizer documents released in May 2022 reveal that despite the large number of adverse events and brand-new health problems catalogued in the case reports —"often serious and often requiring the hospitalization of the patients involved"—the company classified nearly all as "unrelated" to its vaccine.[138,139] Among the acute health events dismissed as "unrelated" were a wide range of serious heart and respiratory problems as well as sudden emergencies such appendicitis and small bowel obstruction.

PRIMARY "SYSTEM ORGAN CLASSES" AFFECTED BY PFIZER SHOTS

Pfizer's clinical trial data pointed to effects on a wide range of "system organ classes":

- General disorders and administration site conditions (32% of adverse events)
- Nervous system disorders (16%)
- Musculoskeletal and connective tissue disorders (11%)
- Gastrointestinal disorders (9%)

- Respiratory, thoracic, and mediastinal disorders (6%)
- Skin and subcutaneous tissue disorders (5%)
- Injury, poisoning, and procedural complications (4%)
- Infections and infestations (3%)
- Investigations (2%)

Despite these far-reaching effects, the only safety concern that Pfizer saw fit to identify as an "important identified risk" was anaphylaxis. It also told the FDA that information on "use in pregnancy and lactation" and use in children under age 12 was "missing."

Source:
https://childrenshealthdefense.org/wp-content/uploads/pfizer-doc-5.3.6-postmarketing-experience.pdf

The Pfizer report detailing three months of clinical trial data also documented 1,223 fatalities. Analyzing data from the second half of 2021 on excess deaths in the millennial generation (approximately ages 25–44), investment advisor Edward Dowd has pointed out that the number already exceeds Vietnam War casualties.[140] Dowd considers the data—and the absence of other plausible explanations—to be a "smoking gun that the [COVID] vaccines are causing excess mortality."

Even before the rollout of COVID vaccines, health insurers were issuing warnings about the growing number of millennials with chronic illness, reporting double-digit increases in the prevalence of some chronic conditions compared to the preceding generation (Generation X).[141] This disturbing trend may be explainable by the fact that the millennial generation partially coincides with the time period in which the childhood vaccine schedule ballooned. COVID injections are now making the situation worse.

Early on, whistleblower Brook Jackson called into question the integrity of Pfizer's vaccine clinical trial for adults, telling *The BMJ* that "the company falsified data, unblinded patients, employed inadequately trained vaccinators, and was slow to follow up on adverse events."[142] Attorney Robert Barnes now represents Jackson in a false claims lawsuit against Pfizer,[143] in which Jackson alleges that the trial was "riddled not only with error but with fraudulent and false certifications to the U.S. government." In response, Pfizer moved to dismiss the case, adopting an unusual legal strategy. Barnes explained:

> *"[T]heir grounds to dismiss. . . is that it doesn't matter if they submitted fraudulent certifications to the government; it doesn't matter if they submitted false statements under penalty of perjury to the government; it doesn't matter if they lied about the safety and efficacy of these drugs—mislabeled, in my opinion, as 'vaccines'—because the government was in on it with them! The government knows what's going on and the government still would have given them a check anyway, so is it really fraud if the government's their co-conspirator? That is, in essence, Pfizer's defense. . . ."*[144]

British pathologist Clare Craig has similar criticisms of Pfizer's clinical trial for babies; describing shocking misrepresentations and omissions of data, Craig argues that "the trial should be deemed null and void" and says, "Parents should be demanding that the decision makers [at FDA] explain themselves."[145]

A FLOOD OF INJURIES

By May 13, 2022, three experimental COVID shots (those made by Pfizer, Moderna, and J&J) had added nearly 1.3 million adverse event reports to VAERS—two-thirds submitted domestically and one-third sent in by manufacturers' foreign subsidiaries—including close to 28,000 new deaths.[146] Of the almost 13,000 COVID-vaccine-related deaths reported on U.S. shores, one-fifth of the individuals perished within 48 hours of vaccination.

For the COVID shots, CDC also inaugurated an additional vaccine safety surveillance system—a smartphone app called v-safe that reportedly allows vaccine recipients to "quickly and easily share with CDC how [they] feel after getting a COVID-19 vaccine."[147] CDC claims the app is helping it monitor safety "in near real time." However, despite three

Freedom of Information Act (FOIA) requests by the Informed Consent Action Network (ICAN), CDC has refused to release 2021's v-safe data to the public; ICAN is suing CDC and HHS for "improperly withholding" vital data from the American people.[148]

While keeping the v-safe data close to the chest, CDC researchers published a report in August 2021 that included some revealing tidbits. The study covered adverse reactions experienced by 129,000 v-safe-enrolled adolescents (in the 12–15 and 16–17 age groups) who received Pfizer COVID shots between December 14, 2020 and July 16, 2021.[149] The CDC authors disclosed the following:

- About half (49%) of adolescents aged 12–15 years reported systemic reactions within a week of receiving dose 1; within a week of dose 2, this percentage increased to more than three in five (63%).
- For teens aged 16–17 years, 56% reported systemic reactions within a week of dose 1, and 70% did so after dose 2.
- The day after the second dose, nearly one in four teens were "unable to perform normal daily activities."

CDC measured "systemic" vaccine reactions—which it defined as abdominal pain, chills, diarrhea, fatigue, fever, headache, joint pain, muscle pain, nausea, rash, or vomiting—for only seven days following each dose, and dismissed this group of symptoms as "mostly mild" or "moderate." In many cases of COVID vaccine injury, however, systemic reactions of this type have proven to be the precursors to far worse. For example, a healthy 21-year-old pre-med student who received a single Pfizer injection started out with nausea and vomiting, but then went on to develop "back pain, a severe rash, tinnitus, eye problems, kidney failure, deafness and neuropathy," as well as enduring lengthy hospitalization and multiple surgeries.[150]

Data on Moderna's clinical trial participants (for the 18–64 age group) showed even higher levels of systemic reactions—82% after dose 2.[151]

"MUCH POTENTIAL FOR HARM"

In April 2021, CDC used v-safe data collected from pregnant women to make the claim that there were no "obvious safety signals" associated with mRNA COVID vaccination during pregnancy.[152] However, corrected analyses of CDC's data by independent European and other researchers showed that 82% of women who received COVID vaccinations prior to week 20 of their pregnancy experienced miscarriages—an "alarming incidence" three to eight times higher than the background rate of second-trimester miscarriage.[153] As for VAERS, by April 1, 2022, the system was receiving an average of 251 fetal death reports a month following COVID vaccination (4,023 fetal deaths reported over a 16-month period), versus an average of 1.5 monthly fetal deaths reported for flu shots over the previous three decades.[154] Military whistleblowers have described a 300% spike in miscarriages in 2021 over the previous five-year average.[155]

In addition to the pregnancy risks, women all over the world are reporting significant menstrual cycle changes following receipt of COVID vaccines—in some instances leading to hysterectomies, and in other cases raising serious questions about future fertility.[156] Nor are the concerns about fertility limited to women. A study published in mid-2022 highlighted large decreases in vaccinated men's sperm counts after a second dose of the Pfizer shot.[157]

In September 2021, a team of scientists from Greece, Italy, Romania, Russia, and the U.S. pushed back against calls to give children COVID vaccines in a paper in *Toxicology Reports* titled, "Why are we vaccinating children against COVID-19?"[158] Emphasizing that COVID deaths in children are "negligible," while the number of post-vaccine deaths reported to VAERS is "high," the paper's authors made the following points:

- Clinical trials "did not address long-term effects most relevant to children."
- Pfizer and regulators ignored "early warning indicator issues," and this neglect "translated into disastrous consequences during the mass inoculation rollout."

- Early COVID-vaccine-related adverse reactions reported to VAERS for the 0-17 age group (through mid-June 2021) provided indications of damage affecting the cardiovascular, gastrointestinal, neural, immune, and endocrine systems as well as vision and breathing problems.
- With potential impacts on all major body systems and major organs, "any mid- or long-term adverse events that emerge could impact children adversely for decades." Moreover, "adverse effects may be cumulative and irreversible, and therefore injury and death rates may increase with every additional inoculation."

Concluding that there is "much potential for harm from the inoculations" to both the elderly population most at risk and the younger population not at risk, the authors voiced the doubts that many others around the world have been expressing:

> *"[I]t is unclear why this mass inoculation for all groups is being done, being allowed, and being promoted."*

Perhaps because this group of authors pulled no punches, the publisher (Elsevier) went on to retract the article in 2022. Elsevier is a subsidiary of the London-based multinational RELX, whose institutional investors (including global asset managers BlackRock and Vanguard) own over 50% of RELX and together "can probably strongly influence board decisions."[159]

For further reading, see these Children's Health Defense e-books:

- *COVID-19 Treatment and Vaccine Decisions from a Pediatric Perspective*
- *Protecting Individual Rights in the Era of COVID-19*

CHAPTER TWO

The High Cost of Vaccine Injuries

HIT-AND-MISS COMPENSATION FOR THE INJURED

In lieu of lawsuits, the NCVIA created the taxpayer-funded National Vaccine Injury Compensation Program (NVICP), an administrative mechanism that opened for business in 1988. The program allows individuals—at least those who know about it and whose injury falls within the narrow three-year statute of limitations[160]—to seek financial redress for some vaccine injuries. Congress designed the NVICP primarily for injuries experienced by children, but in recent years, compensation—when awarded at all—increasingly has gone to adults[161] injured by seasonal flu shots.[162]

Although it is not necessary to have an attorney file a claim, less than 1% of petitions submitted to NVICP without an attorney's help achieve success.[163] Even with the assistance of an attorney, petitioners' claims are, more often than not, filed in vain; two-thirds[164] of claims either get dismissed or languish on endless hold in the backlogged system.[165] To the third of petitioners thus far lucky enough to have obtained some compensation, the taxpayer-funded program paid out $4.7 billion between 1988 and early 2022.

UNCOMPENSATED COUNTERMEASURES

In fiscal year 2010, HHS established a separate program, the Countermeasures Injury Compensation Program (CICP),[166] which theoretically is available to award minimal compensation for serious injuries or deaths resulting from EUA countermeasures, including EUA vaccines. However, the CICP has even more stringent conditions than the NVICP—a one-year statute of limitations, no coverage of legal fees, and no compensation for pain and suffering.

As of March 2022,[167] the underfunded CICP—notorious for its "cumbersome claims process and low likelihood of success for claimants"[168]—had compensated only 30 non-COVID petitioners, representing 0.4% of all claims filed since 2010. Claims related to COVID countermeasures—none of which, to date, have been compensated—now constitute 93% of total claims filed (see "NVICP and CICP Statistics").

An emeritus law school professor quoted in an April 2022 article in *The BMJ* described the CICP as a "horrible" program, noting that claims are "dealt with secretly," with "such a lack of transparency. . . that it's frightening."[169] The article—titled "Covid-19: Is the US compensation scheme for vaccine injuries fit for purpose?"—also noted that the CICP's burden of proof is virtually impossible to meet. Others in the legal profession agree. As reported by Reuters, a firm experienced in vaccine-related injury claims is telling the hundreds of individuals injured by COVID injections who have contacted them, "Our law firm has concluded that there is nothing our attorneys can do to significantly assist you."[170] Or, as another attorney quoted by Reuters put it, "COVID vaccine claimants have two rights: 'You have the right to file . . . . And you have the right to lose." In other words, "if you've suffered an injury related to the Pfizer, Moderna or Johnson & Johnson vaccines, you're basically out of luck."

The lack of legal recourse available to the vaccine-injured—and especially to the COVID-vaccine-injured—is so comprehensive and so egregious that even complacent news outlets like CNBC apparently felt compelled to say something on the occasion of the experimental injections' rollout in December 2020. In an article titled "You can't sue Pfizer or Moderna if you have severe Covid vaccine side effects. The government likely won't compensate you for damages either," CNBC explained:

> *"If you experience severe side effects after getting a Covid vaccine, lawyers tell CNBC there is basically no one to blame in a U.S. court of law. The federal government has granted companies like Pfizer and Moderna immunity from liability if something unintentionally goes wrong with their vaccines . . . . You also can't sue the Food and Drug Administration for authorizing a vaccine for emergency use . . . ."*[171]

HOLDING THE BAG FINANCIALLY

Given the seemingly intentional deficiencies of the government's two compensation programs, the parents of vaccine-injured children—and individuals injured as adults—sooner or later discover that they are on their own for expenses associated with the injury. The costs can be substantial and long-lasting. They include health care expenses not covered by insurance, such as out-of-pocket "alternative" therapies (that is, if the person or family even has insurance to begin with); physical or occupational therapy; educational or training support; in-home care; wheelchairs; supplements; special foods; and much more. Moreover, such expenses often arise in the context of reduced income—either because a parent has to stop working to care for an injured child or because an injured adult is no longer able to work full-time or at all.

FINANCIAL TRAIN WRECK: LESSONS FROM AUTISM

Among the range of adverse outcomes experienced by vaccine-injured Americans, autism is one of the more costly,[172] as several of the stories included in this book illustrate. Parents of children with ASD, and particularly autistic children with intellectual disabilities, face lifetime costs that, as of 2013, were estimated at around $2.4 million—an amount that could well be higher a decade later.[173] At the national level, researchers Toby Rogers, Mark Blaxill, and Cynthia Nevison predict that autism costs in the U.S. will reach over half a trillion by 2030 ($589 billion) and

National Vaccine Injury Compensation Program
(Oct. 1, 1988–Mar. 1, 2022):

- Total claims filed: 24,824
- Total claims adjudicated: 20,797 (84%)
- Number of claims dismissed: 12,030 (58% of adjudicated)
- Unadjudicated claims: 4,027 (16%)
- Number of claims compensated: 8,767 (35% of filed claims)
- Total compensation awarded: $4.7 billion
- Average compensation: $536,101

Countermeasures Injury Compensation Program
(FY2010–Mar. 1, 2022):

- Total claims filed: 7,547*
- Claims deemed eligible for review: 7,454
- Claims "pending review or in review": 7,049
- Claims denied or dismissed: 457 (6% of filed claims)
- Claims compensated—H1N1 or smallpox vaccines: 30 (0.4% of filed claims)
- Claims compensated—COVID-19 vaccines: 0

**Includes 7,056 claims (93% of total) filed for injuries or deaths resulting from COVID-19 countermeasures—including 4,097 COVID-19 vaccine claims and 2,959 claims for "other COVID-19 countermeasures" (primarily ventilator intervention and remdesivir). The remaining 491 claims were filed for injuries or deaths resulting from H1N1, smallpox, anthrax, or other vaccines or countermeasures.*

Sources:

Data & Statistics. Health Resources & Services Administration, updated March 1, 2022. https://www.hrsa.gov/sites/default/files/hrsa/vaccine-compensation/data/vicp-stats-03-01-22.pdf; Countermeasures Injury Compensation Program (CICP) Data: Aggregate Data as of March 1, 2022. Health Resources & Services Administration. https://www.hrsa.gov/cicp/cicp-data.

$5.54 trillion by 2060,[174] with lost productivity and adult care being two of the most significant contributors.

This financial train wreck was not hard to see coming. Back in 2007, a researcher was already warning:

> *"The substantial costs resulting from adult care and lost productivity of both individuals with autism and their parents have important implications for those aging members of the baby boom generation approaching retirement, including large financial burdens affecting not only those families but also potentially society in general."*[175]

Many autism parents have learned the hard way about the impact that a severe vaccine injury can have on a parent's ability to work outside the home.[176] Studies show that for mothers, having an autistic child significantly lowers the number of hours worked per week and the number of months worked per year, with a similar but smaller effect for fathers.[177] Australian researchers report that mothers of school-age autistic children have "up to two times the odds" of not being in the labor force compared to other mothers, even though parental employment may be needed to "ensur[e] financial ability to access care."[178] As one autism website explains, autism entails "additional expenditures that can turn a middle-income family into a low-income family in a matter of months"[179] (see "What Autism Parents Say").

Autism researchers agree that "ASD affects the lives of all family members, implying the beginning of a difficult life for them" and often including financial struggles.[180] As a 2010 study in the *Journal of Family Psychology* reported, this "extraordinary level of stress" not infrequently also "take[s] a toll on marriages," which has further implications for family finances.[181] In that study—which, strangely, avoided discussing autism's financial impacts—the researchers compared divorce rates for parents of an adolescent or adult child with autism to the rate for parents of comparably aged children without a disability. They found that at least one in four parents (24%) of an autistic child divorced—with a "prolonged period of vulnerability to divorce" that persisted until the child was a 30-year-old adult—whereas 14% of the comparison group divorced, with the risk of marital dissolution peaking around the child's eighth year.

Anecdotal evidence suggests a much higher autism-related divorce rate—possibly as high as 80%—although those eager to downplay autism's impacts contest the higher estimates. Divorce attorneys state, "Regardless of the statistics, one thing does not change—caring for an ASD child is difficult and it often places strain on marriages," in part because parents "will try to do everything they can to help their child," even when it leads to financial disaster.[182] Doing "everything they can" may include "skipping meals to be able to afford therapy for their children" and "robbing their future by depleting savings, emptying their 401K plans, selling stocks and even filing for bankruptcy."[183]

What Autism Parents Say

Parents posting comments on the *Age of Autism* website a decade ago had the following to say about the financial stresses faced by many autism families:

- Posted by "Kym": "Not only is it super costly to raise a child with autism with all the therapies but often one parent has to quit working. I know I tried to keep working but it was so hard. . . kept having to take off for doctors appointments and to go up to school. I ended up having to quit my full time job to take care of my son and we took a big financial hit. And if you can manage to somehow keep working, what happens when our kids get to 21 and there are no services? No more school. Someone has to stay with the adult if there are no services. It is a bad situation."
- Posted by "Fed Up": "[M]y nonverbal 11 year old 140 lb child will continue to drain $60K a year b/c insurance covers nothing but at least we can afford $30K a year for ABA [Applied Behavior Analysis] and live pay check to pay check no longer

contributing to any retirement or. . . medical savings. What a great life."

- Posted by "Sherry": "I know people who divorce because then their income gets lowered and the child is able to get the services he/she needs because they are (finally) considered 'Low Income.' They can get help with: Medical, Dental, Vision, HBTS [Home-Based Therapeutic Services], Occupational Therapy and other therapies. Also: Food Stamps, Heating Assistance, help with Electric bills, and many other bills. ...[S]ometimes people have to 'cheat the system that cheats the kids' in order to get the help they need."

Source:

Do couples divorce because of autism? *Age of Autism*, Mar. 20, 2011. https://www.ageofautism.com/2011/03/do-couple-divorce-because-of-autism.html

ANOTHER TRAIN WRECK

Labor force participation has also become impossible for many adults severely injured by COVID shots.[184] After a single dose, for example, a pilot lost both his health and career, leaving his family with an income "20% of what it was before" and facing "mounting debt and unpaid taxes."[185]

Nor can COVID-vaccine-injured individuals who hang on to their jobs be assured of getting workers' compensation, with widespread reports of denied claims.[186] As of March 2021, employment law experts described this as "a developing area of the law," suggesting that workers' compensation for adverse reactions to COVID injections might be forthcoming only in "certain scenarios," hinging on factors such as state laws, whether the employer mandated the jab or the employee took the shot "voluntarily," and whether the injury is determined to be "work-related."[187] Some attorneys believe that employers who provide COVID shots on-site qualify for the PREP Act's liability protections in their role as "program planners."

On the other hand, a vaccine liability bill in the state of Missouri provides a legal framework for employees to hold their employers liable if they experience an adverse event from an employer-mandated COVID shot.[188]

MEDICAL DEBT AND VACCINE INJURY

Medical debt bears responsibility for a sizable proportion of consumer bankruptcies and is estimated to be the primary cause of anywhere from one in four[189] to three in five[190] bankruptcies. In her former capacity as an investment advisor to individuals and families, Catherine Austin Fitts found that many of her clients, subscribers, and their children "had been devastated and drained by health care failures and corruption—and the most common catalyst for this devastation was vaccine death and injury."[191] Moreover, such cases were not unusual. Fitts also saw that the financial costs of vaccine injury had significant implications for the allocation of family resources—including not just money but attention, time, and other assets—affecting not only the injured children but also the parents, grandparents, siblings, and even future generations.

Grandparents, in particular, often are called on or feel compelled to help financially when a grandchild is injured by vaccines, because it is generally the grandparents who have accumulated the most family wealth. The inheritance accumulated by the older generation over a lifetime of hard work and saving—perhaps intended toward tuitions, business startups, or a rainy day fund—instead is depleted by a tragedy that, with proper due diligence, might have been prevented.

Traditionally, informed consent forms for vaccination do not provide disclosure or statistics related to the financial costs associated with injury, disability, or death, nor do they explain an injury's impact on other family resources. When the COVID shots began to be authorized on an emergency basis, Fitts developed a "Family Financial Disclosure Form for COVID-19 Injections" that seeks to rectify this important omission,[192] helping families communicate about the sizable financial risks and identify specific actions that can and should be taken to protect the family from financial devastation and bankruptcy.[193]

IMPLICATIONS FOR HOUSE AND HOME

As of early 2020, families (versus individuals) made up about a third of the U.S. homeless population.[194] A Seattle-based study published around that time, just before the declared COVID "pandemic," found that medical debt played a large role in homelessness.[195] Based on respondents' self-report, the researchers ascertained the following:

- Most homeless respondents (over 80%) had at least one kind of debt—such as medical debt, credit card debt, payday loans, or student loans—but primarily medical debt.
- Two-thirds of those with any debt had current medical debt in amounts ranging from $100 to $50,000.
- Health insurance was "not protective"—more than half of those with medical debt incurred the debt while insured (mostly through Medicaid).
- Almost half had trouble paying medical bills (either their own or family members' bills), even when the amounts were under $300.
- About one-third attributed their homelessness, at least in part, to medical debt.
- Most significantly, medical debt extended homelessness by an average of nearly two years compared to individuals not encumbered by unpaid medical bills.

As pointed out on the website debt.org, COVID-19 has "exacerbated" the interrelated problems of medical debt and housing instability.[196] With inflation, home prices, mortgage rates, and rents all soaring,[197] homeless shelters are warning that many families "are one paycheck away from being homeless."[198] Although debt.org stops short of discussing medical debt triggered by COVID vaccine injuries, it stands to reason that the hundreds of thousands of dollars in medical expenses being incurred by many of the people injured by the experimental shots will have knock-on effects on families' housing security,[199] including increasing the risk of mortgage default and foreclosure.[200] First-mortgage default rates spiked in early 2022,[201] around the same time that reports began surfacing of a dramatic and unprecedented surge in all-cause mortality in working-age adults—a surge coinciding with the administration of COVID shots.[202]

For further reading, see the following Children's Health Defense articles in *The Defender*:

- "Vaccine-induced myocarditis injuring record number of young people. Will shots also bankrupt families?" (*The Defender*, Jan. 31, 2022).
- "Exclusive: 30-year-old still seeking answers 6 months after developing neurological complications following Pfizer vaccine" by Megan Redshaw (*The Defender*, Sep. 8, 2021).
- "Exclusive interview: Mom whose 14-year-old son developed myocarditis after Pfizer vaccine no longer trusts CDC, public health officials" by Megan Redshaw (*The Defender*, Aug. 11, 2021).
- "Woman with 'life-altering' injuries after COVID vaccine teams up with U.S. senators to demand answers" by Megan Redshaw (*The Defender*, Jul. 14, 2021).
- "Hundreds injured by COVID vaccines turn to GoFundMe for help with expenses" by Megan Redshaw (*The Defender*, Jul. 8, 2021).
- "Injured by a COVID vaccine? Want financial compensation? Too bad, says injury compensation law firm" by Megan Redshaw (*The Defender*, Jul. 1, 2021).
- "Woman who nearly died after J&J vaccine stuck with $1 million medical bill, says government should pay" by Megan Redshaw (*The Defender*, Jun. 2, 2021).

CHAPTER THREE

Childhood Vaccine Injuries: Autism Stories

AUTISM—A DECADES-LONG VACCINE INJURY TSUNAMI

Children have suffered vaccine-induced brain damage since the inception of mass vaccination.[203] These brain injuries have been the equivalent of a neon sign flashing urgent warnings about vaccination's dangers. Rather than acknowledge that vaccines can and do cause brain damage, however, the medical community has essentially whited out this and other forms of vaccine injury, conjuring up a multitude of symptom-focused diagnostic labels that function as cover-ups for the undiscussed root cause.

One of these labels, ASD, is an umbrella diagnosis that encompasses a multiplicity of neurological and other symptoms. In the 1970s, the phenomenon now categorized as autism was estimated to affect fewer than 3 in 10,000 children.[204] Since then, its prevalence has surged.[205] Among 3- to 17-year-olds, an estimated 1 in 34 children are affected (2019–2020 National Survey of Children's Health); the most recent CDC data situate autism at 1 in 44 eight-year-olds (as of 2018)—and 1 in 28 boys.

Although environmental toxicants clearly also contribute to ASD—with research highlighting culprits such as air pollutants and glyphosate[206]—the role of vaccines as a potent direct and indirect trigger cannot be ignored. Yet, through gaslighting and what writer Jon Rappoport calls "sophisticated deceptive diversion"—including the "shuffling [of] various disease and disorder labels; studies claiming there is no link between vaccines and autism; the hoops the government makes parents jump through . . . to try to obtain financial compensation . . . ; the legal deal allowing vaccine manufacturers to avoid law suits; [and] the invented cover stories claiming autism begins in utero or is a genetic disorder"—the CDC, other government agencies, and the media largely have managed to bury the vaccine–autism link "in a mountain of obfuscation."[207]

The following stories furnish a partial illustration of what life is like for many children and families who have experienced the form of vaccine damage known as ASD.

For further reading, see the following Children's Health Defense and *Defender* articles:

- "Gaslighting autism families: CDC, media continue to obscure decades of vaccine-related harm" (*The Defender*, Dec. 17, 2021).
- "Autism—the most glaring aspect of the deterioration of health among our kids?" by Anne Dachel (Children's Health Defense, Aug. 21, 2020).
- "Vaccines and autism—is the science really settled?" by JB Handley (Children's Health Defense, Aug. 11, 2020).
- "Measles vaccination and autism: the inexcusable suppression of a long-documented link" (Children's Health Defense, Jul. 9, 2020).
- "Autism spin versus autism trends: rising prevalence in black and hispanic children" (Children's Health Defense, Aug. 29, 2019).
- "The autism blame game: this time, it's the womb" (Children's Health Defense, Jul. 9, 2019).
- "Thousand-fold increase in autism prevalence since the 1930s" (Children's Health Defense, Jul. 10, 2018).

The Autism Merry-Go-Round

Largely left to their own devices by the medical establishment, parents of autistic children have sought out and explored a wide range of therapeutic interventions to help their children detoxify, develop, and recover to the extent possible. Over the years, it has not been unusual for autism families to try some (or all) of the following, with varying degrees of success:

- Acupuncture
- Allergy treatments
- Antibiotics for PANS (pediatric acute-onset neuropsychiatric syndrome), PANDAS (pediatric autoimmune neuropsychiatric disorder associated with streptococcal infections), and tics
- Antifungal and antiviral pharmaceutical intervention
- Chelation (removal of heavy metals) via nutrients, intravenously, and/or with oral penicillimine
- Colonics (to cleanse metals and toxins following chelation)
- Doman sensory inputting
- Food therapy (to develop ability to taste, chew, and swallow different textures)
- Gastrointestinal treatments
- Greenspan Floortime Approach
- Intravenous immunoglobulin (IVIG) to manage immunodeficiency
- Mitochondrial "cocktail" (CoQ10, carnitine, glutathione, and other supplements)
- Nicotine patch (for tics)
- Nutritional interventions (gluten-free/casein-free diet; avoidance of additives/preservatives)
- Nutritional support for methylation
- Occupational therapy
- Play therapy

- Prism glasses (to increase visual field focus)
- Private one-on-one therapy
- Sauna (for detox)
- Secretin
- Supplementation
- Speech therapy
- Steroid treatment (for epilepsy)
- Tomatis method (to enhance listening and communication skills)
- Tutoring
- Weighted vests (to help with proprioceptive feedback and sense of space)

Sources:

Connery K, Tippett M, Delhey LM...Frye RE. Intravenous immunoglobulin for the treatment of autoimmune encephalopathy in children with autism. *Transl Psychiatry*. 2018;8:148.
Doman RJ Jr. Neurodevelopmental perspectives on autism and Asperger's syndrome. *Autism Health and Wellness*. 2009;1(3). Available at: https://www.nacd.org/neurodevelopmental-perspectives-on-autism-and-aspergers-syndrome/
Frye RE, Rossignol D. Mitochondrial physiology and autism spectrum disorder. *OA Autism*. 2013;1(1):5.
Greenspan Floortime Approach. https://www.stanleygreenspan.com/
Kresser C. Methylation and autism—is there a connection? Apr. 18, 2019. https://chriskresser.com/methylation-and-autism/#Testing_and_Treating_Methylation_Problems
Madra M, Ringel R, Margolis KG. Gastrointestinal issues and autism spectrum disorder. *Child Adolesc Psychiatr Clin N Am*. 2020;29(3):501-513.
Shea C. Benefits of weighted vests for children with autism. NAPA Center, Jun. 3, 2020.
Sinclair DB. Prednisone therapy in pediatric epilepsy. *Pediatr Neurol*. 2003;28(3):194-198.
Tomatis method. https://www.tomatis.com/en

Temple's Story

[*As told by Temple's mother, Dr. Sheila Ealey, on March 23, 2022. Dr. Ealey serves on the CHD Board of Directors.*]

The Overview

At almost 13 months of age, in August 2000, Temple Ealey received four simultaneous vaccines that altered the course of his life forever, landing him with a diagnosis of severe autism six months later. At the time, Temple's father was a United States Coast Guard officer, and the family (including Temple's twin sister and an older sister) resided on a Maryland naval base, where they received their medical care. Following many years of efforts by his parents to heal him, Temple, now age 22, is happy but has the developmental capacity of a six-year-old.

Warning Signs

Prior to his fateful 12-month appointment, Temple had received two earlier rounds of vaccines at the naval base: a hepatitis B shot at birth; and, at five months of age, polio, *Haemophilus influenzae* type b (Hib), and diphtheria-tetanus-acellular pertussis (DTaP) shots.

Sheila, Temple's mother, had requested that Temple not get the birth dose of hepatitis B, but the pediatrician, claiming to have "forgotten" this request, administered it against her wishes. Temple then developed jaundice—a hallmark symptom of hepatitis B. Jaundice shows up as a vaccine-related adverse event in hundreds of VAERS reports[208] and as a documented adverse event associated with prenatal hepatitis A and B vaccination.[209] Jaundice[210] and autoimmune hepatitis[211] have also been increasingly reported in conjunction with COVID vaccination.[212]

After his five-month shots—three shots featuring five different antigens—Temple developed an extremely high fever, a warning sign that his body was having trouble coping with the immune onslaught.[213] At this juncture, Temple's mother Sheila began to suspect that there was "something wrong with these vaccines," but the military doctors dismissed her concerns, telling her that Temple's body was just "adjusting" and advising her to give him acetaminophen (Tylenol). Much later, she learned about an extensive body of research that identifies acetaminophen as both an independent and synergistic risk factor for autism.[214]

Around this time, Temple also experienced sudden atrophy of one of his testicles, an abnormality that prompted a referral to a pediatric urologist. At a loss as to causation, the specialist—apparently unaware that some DTaP inserts[215] list testicular atrophy as a serious adverse event—merely stated, "We'll just have to watch this."

The Tipping Point

Following the five-month appointment, the naval base clinic badgered the family to bring Temple and his twin back for more shots. Sheila delayed their return until they were 12 (almost 13) months old. Temple then not only received further DTaP and Hib shots but was also given a double dose of MMR vaccine—one dose intended for him and a second laid out on the table for his twin sister.

Alarmed by the double-dose error, Sheila immediately departed with both children, a hidden blessing for her daughter, who was thereby spared receiving any 12-month shots at all. The only shot Temple's twin received at five months was a DTaP shot, following which she developed terrible post-vaccination gastroesophageal reflux. Adverse gastrointestinal reactions are listed in numerous vaccine package inserts.[216]

Back at home, Temple's reaction to the shots was immediate. He started "banging his head against the wall, the floor, anything he could"—a sign that "his brain was on fire." Although he had already learned to walk and was walking well, he reverted to crawling—backwards. After crying the entire night, by the next morning, he was staring into space and could not hold on to his mother when she picked him up. From this point on, the Ealeys were "done" with vaccinating.

The Diagnosis

Almost immediately, Sheila left Maryland and returned to her roots in New Orleans, where she had previously established a relationship with a more trusted pediatrician. The pediatrician instantly theorized that Temple had autism. Until Temple's injury, the Ealeys had scarcely even heard of autism and had never encountered it in any African-American family of their acquaintance. (Sheila recalls that when she saw the movie *Rain Man*, released in the late 1980s, she assumed the condition was a Hollywood fabrication.)

In New Orleans, Sheila obtained an appointment with a developmental neurologist. Before Temple and his father entered the room, Sheila had an opportunity to speak with the specialist, who began proposing a lengthy list of possible tests, indicating that they should explore a range of diagnoses because autism was "very rare." When Temple arrived, however, the neurologist "looked astonished and said, 'I do not need to do one single test on this child. This child is profoundly autistic.'" Referencing Leo Kanner—the child psychiatrist who in the 1940s published the first descriptions of children with early "infantile autism"—the neurologist pronounced Temple's condition "true Kanner autism."

Learning that an autism clinic was about to open in New Orleans, Sheila called the new clinic's director. When he said the clinic would not open for another month, Sheila answered, "my son cannot wait a month"; the director ended up agreeing to see them the following week. Thereafter, they made the rounds between occupational therapists, development pediatricians, psychiatrists, psychologists—"you name it"—who continued to tell them that Temple was among the "rare and worst cases" of autism they had ever seen. By this time, the Ealeys were fully convinced "it was the vaccines."

Much later, Temple was diagnosed with a primary mitochondrial disease (PMD). Although PMDs are considered to be inherited, mitochondrial experts note that "the impact of environmental stressors can contribute" to the genetic mutations observed in PMD.[217] In many individuals, this type of mutation is dormant and inconsequential; in Temple's case, "the vaccines pulled the trigger."

Medical Experimentation

In a 2014 interview,[218] Sheila noted a "higher number of autistic children in the military." Some military families believe that the disproportionate rate of autism observed in military populations can be attributed, at least in part, to the "large amount of vaccines military parents receive, above and beyond those given to the civilian population,"[219] including anthrax vaccines.[220]

Until Temple's injury, Sheila had believed "a doctor wouldn't hurt my child." As she put it in the 2014 interview, educated African-Americans such as her and her husband "forgot about the many medical atrocities we learned of from our grandparents about how never to trust doctors, because of the Tuskegee experiment, etc."[221] One such atrocity had even directly affected her grandmother, a story Sheila recounts in the film, *Medical Racism*.[222] During one of her grandmother's pregnancies, her grandmother later told her, the "white coats" came and administered "pox shots" to eight pregnant women living and working on a cotton plantation; all eight, including Sheila's grandmother, ended up having children "afflicted" with Down syndrome or other birth defects (see "Autism and Birth Defects"). "No one had ever seen such a thing before," Sheila notes. In all, her grandmother had 22 children, including multiple sets of twins, all of whom—with the exception of the "afflicted" child—were healthy.

Sheila herself received no vaccines during childhood and consequently, "knew nothing of vaccines." With parents who had grown up on the land and who had "a remedy for everything," there was no need to go to doctors. Ealey and her siblings experienced childhood infections like chickenpox, rubella, and scarlet fever naturally and uneventfully. On the rare occasions when someone in the family needed outside care, they saw chiropractors and paid out of pocket. When Sheila got married, however, she was instructed to get a tetanus shot and recalls getting sick afterwards.

Day-to-Day Health Impacts

Daily life with Temple included "failure to thrive, explosive diarrhea, and doctor's appointment after doctor's appointment." The Ealeys also rapidly discovered that "there's no such thing as a balanced life when you have a child who is profoundly affected, especially for the siblings."

In the early 2000s in New Orleans, moreover, "there was nowhere to take a child" as severely autistic as Temple. Up to that point, Sheila's professional training had been in communications and marketing, not education, but—being solutions-focused—she began learning the therapeutic approach known as applied behavior analysis (ABA).[223] At the time, ABA was the "only treatment. . . supported by substantial empirical research" for individuals with autism, and one of the few therapies that stood any chance of being covered by insurance. (Sheila notes that later, she learned of preferable therapies such as the Greenspan Floortime Approach,[224] but insurance would not cover it.)

Autism and Birth Defects

In a study published in the mid-1970s, a German researcher linked postnatal smallpox vaccination to infantile autism, and in 2017, a CDC analysis of VAERS data reported a link between prenatal vaccination and birth defects. The CDC researchers noted that while major birth defects were "important infant outcomes," they had "not been well studied in the post-marketing surveillance of vaccines given to pregnant women." They also hypothesized significant underreporting of birth defects to VAERS, "not only because of the spontaneous nature of VAERS, but also due to the period of time between vaccination and delivery, and the fact that many defects are not necessarily obvious or symptomatic immediately after birth."

Sources:

Eggers C. [Autistic syndrome (Kanner) and vaccination against smallpox (author's transl)]. [Article in German] *Klin Padiatr.* 1976;188(2):172-180. https://pubmed.ncbi.nlm.nih.gov/944354/

Moro PL, Cragan J, Lewis P, Sukumaran L. Major birth defects after vaccination reported to the Vaccine Adverse Event Reporting System (VAERS), 1990–2014. *Birth Defects Res.* 2017;109(13):1057-1062. https://pubmed.ncbi.nlm.nih.gov/28762675/

After meeting several affluent families with autism-affected children—and noticing, in passing, that "when you have this diagnosis, it has no respect of person"—she began hiring herself out to provide ABA services to other autistic children to support her ability to help Temple. Eventually, she founded and directed a small therapeutic day school for children with severe ASD and other nonverbal intellectual disabilities, along the way acquiring master's and doctoral degrees in special education.

At some point, Sheila learned of the Defeat Autism Now! (DAN!) project launched in the mid-1990s by the Autism Research Institute (ARI)[225]—the institute founded in 1967 by Dr. Bernard Rimland to research autism causes and treatments. Through this network, she learned of various biomedical treatments being used by DAN! doctors (physicians who were willing to acknowledge a vaccine-autism link and the problems caused by mercury in vaccines). Although DAN! protocols helped some children recover from autism, and there was considerable enthusiasm at the time about the potential for autism recovery, Sheila now says, "No one told me that with the majority of vaccine injuries, especially when affecting the brain, they may *not* be recoverable." At a conference held in early 2022, a long-time autism doctor admitted to Sheila that a lot had changed since the early DAN! days, stating "we did the best we could, but we didn't know then what we know now." Sheila advises other autism parents about the importance of asking questions and trying a "combination of things, different modalities, because there is no 'one size fits all' with autism."

Financial Impacts

Twenty-one-plus years into her son's vaccine injury, Sheila has learned that "you are virtually on your own, and you have to fight for every little tidbit your child gets." The Ealeys discovered early on that insurance would only cover things like blood tests, other tests (e.g., MRIs, sonograms, electroencephalograms, electrocardiograms), psychotropic drugs, and interventions such as ABA, but would not cover the biomedical interventions or integrative therapies—for which "you have to pay through the nose"—that actually helped Temple in a meaningful way. To remain free of insurance industry constraints, integrative doctors who truly understand autism generally do not take insurance; even so, they often become the targets of ginned-up medical board assaults to pull their licenses.[226]

Dr. Paul Thomas is just such a pediatrician, who had his medical license suspended after he published data showing that unvaccinated children in his practice were far healthier than vaccinated children.[227] Thomas has described how "Reimbursement incentives for vaccines and 'quality measures' contract systems and Pharma-friendly state medical boards effectively force most pediatricians to follow the CDC schedule or risk their practices."[228] Most do so willingly,[229] for, as a private-practice physician affiliated with the CDC has written, childhood vaccines not only furnish "steady revenue" but can also improve a practice's "financial viability" and bottom line.[230]

Sheila, for her part, advises parents not to let ignorance and fear hijack their parental instincts and responsibilities. She asks, "Are you going to trust a doctor who is going to get a bonus at the end of the year for all the kids he vaccinated and is going to live a fabulous life, while you go off with your child and the only options mainstream medicine offers are Tylenol or psychotropic drugs?"

Autism is very expensive, Sheila says: "You need disposable cash, and a lot of it." She conservatively estimates that severe autism costs $150,000-plus per year out of pocket. She notes that she and her husband should be financially comfortable at their age, but instead they cashed out their retirement and downsized their home. "We should have a certain amount of wealth, but we don't because it all went to Temple's injury. The majority of your money will go to trying to fix these issues."

Without the Ealeys' support, Temple would currently be expected to live on $763 per month in Social Security Disability. Medicaid also offers waiver programs for disabled children and young adults[231] that function as "a gateway to getting services,"[232] but Temple has been on a Texas waiting list for years. A Texas agency observes that 15-year wait times for waivers are not uncommon. Waivers, administered at the state level, "let states use Medicaid funds for long-term home and community-based services for people with disabilities or special health care needs in order to help them live in the community."

When Temple was first injured, it not only dramatically changed the family's daily routine but also affected Sheila's in-home availability to her children. She had wanted to homeschool but no longer had the ability to be a stay-at-home mother. Temple's father's workload changed, too; despite being a Coast Guard officer, he had to take on extra work to pay for the in-home support that the family needed.

In the early 2000s—when autism prevalence was estimated to affect one in 150 children—the NVICP consolidated 5,400 compensation claims into something called the Omnibus Autism Proceeding (OAP).[233] The claims were filed by parents, including the Ealeys, who asserted that vaccines had injured their children, causing seizures, developmental delays, and mitochondrial injuries that ultimately led to their child's diagnosis of autism. The NVICP told the thousands of families that it would make a determination about compensation based on six "test cases," using the half a dozen cases to evaluate three narrowly defined theories of autism causation via vaccine injury. Through crooked, stack-the-deck means—characterized by CHD chairman Robert F. Kennedy, Jr. as "one of the most consequential frauds, arguably in human history"—the NVICP dismissed all 5,000-plus petitions (see "OAP Misconduct").

In addition to the fruitless attempt to obtain compensation through the OAP, the Ealeys were involved in a 14-year private lawsuit against Merck (manufacturer of the MMR), seeking to exploit a legal loophole. However, the lawsuit was dismissed after the U.S. Supreme Court issued

OAP Misconduct

Under the provisions of the National Vaccine Injury Compensation Program, vaccine-injured individuals file claims against the HHS secretary in the U.S. Court of Federal Claims Office of Special Masters. The adversarial process pits petitioners not just against special masters who adjudicate the claims but also against U.S. Department of Justice (DOJ) attorneys who "defend HHS."

Had the OAP pinpointed vaccination as the probable culprit in even one of the six "test cases," the compensation program could have been on the hook to compensate all 5,400 families—an outcome that would have bankrupted the program and cast a black cloud over the entire childhood vaccination program. The special masters and the DOJ, therefore, pulled a couple of fast ones.

First, HHS quietly removed one of the test cases; while awarding millions and admitting vaccines were responsible for the child's autism, HHS sealed the documents so the case "could not be used to establish precedent on any of the other OAP cases." Next, two DOJ attorneys allegedly distorted the views of HHS's star expert witness, Dr.

Andrew Zimmerman, who had rejected the vaccine–autism theory of causation in one specific case but had also told the DOJ attorneys that he believed vaccines could indeed cause autism in some children. Years later, in 2019, Zimmerman signed an affidavit disclosing his statements to the DOJ attorneys during the OAP deliberations to the effect that his opinion in the one case was not intended "to be a blanket statement as to all children and all medical science."

Journalist Sharyl Attkisson later noted that Zimmerman's consequential scientific opinion "stood to change everything about the vaccine–autism debate—if people were to find out"—so the DOJ instead fired Zimmerman as an expert witness. Even worse, the two DOJ attorneys intentionally misused Zimmerman's statements, misrepresenting his broader views and omitting the expert's statement that vaccines can and do cause autism in a subset of children.

Source:

Children's Health Defense. Gaslighting autism families: CDC, media continue to obscure decades of vaccine-related harm. *The Defender*, Dec. 17, 2021.

a ruling in 2011, in *Bruesewitz v. Wyeth*, asserting that the NCVIA "preempts all design-defect claims against vaccine manufacturers brought by plaintiffs who seek compensation for injury or death caused by vaccine side effects." This legal verdict was facilitated by the DOJ's and HHS's willful concealment and misrepresentation of evidence during the OAP; as Kennedy and OAP parent Rolf Hazlehurst wrote in 2018, "The same DOJ attorneys subsequently intentionally misled the United States Court of Appeals for the Federal Circuit," and as a result, "fraud was ultimately perpetrated upon the Supreme Court of the United States."[234]

Sheila describes the 1986 Act and the vaccine industry's freedom from liability as "a runaway train." She observes, "Congress didn't think to protect the people; instead, it protected the corporations." Meanwhile, individuals and families continue to discover that "there is a price to pay if you choose to sign up to be a human guinea pig."

Social Impacts

Over the years, the Ealeys have lost family and friends who thought they were "crazy" to blame vaccines for Temple's injury. As Sheila describes it, "Friends literally walked away. They didn't understand the diagnosis." Today, they also have relatives who have taken the COVID shots and still do not want to believe that vaccines can cause physical and financial harm.

To explain the reluctance to question vaccination's safety, Sheila faults the educational system, which no longer offers "decent science and biology classes" yet has "made science into a deity," as well as cognitive dissonance and years of "safe and effective" brainwashing. She notes that with fellow Christians, she has changed how she approaches the conversation, focusing more on the biblical aspects of the vaccination debate. As she advises:

> "*Value yourself and your child; trust the life that God has allowed you to bring into this world. In God's universe, there is no place for fear. Courage will get you through.*"

Temple Today

Sheila credits biomedical interventions with helping Temple to "come such a long way," including being able to speak in complete sentences. However, while Temple "is happy and has his routines, he will never marry and will never have a family. He is not experiencing college life; he should be in grad school and should have girlfriends." Sheila continues to wrestle with "a certain amount of remorse and maternal guilt," stating that "As a Christian, for me the inner voice is the Holy Spirit, and I didn't listen to it." The Ealeys are also only too aware of their own mortality and worry about Temple's care when they're gone.

Kevin's Story

[*As told by Kevin's mother, Karen McDonough, on April 15, 2022. Karen serves as liaison for CHD's U.S. and international chapters.*]

The Overview

Kevin McDonough, currently 27 years old, is the oldest of three siblings and lives at home with his parents in Illinois. Due to significant mercury toxicity from thimerosal-containing vaccines he received as an infant and toddler—and thimerosal-containing biologics given to his mother during her pregnancy—Kevin was diagnosed at age three with pervasive developmental disorder–not otherwise specified (PDD-NOS), a subtype of autism. (In 2013, PDD-NOS was folded into the revised definition of ASD.) After nearly three decades of trying to help Kevin heal, his parents still vow that they will "leave no stone unturned." The tough lessons the McDonoughs learned as a result of Kevin's injury likely spared their younger son—now 25 years old and completing his training as a physician's assistant—and their 18-year-old daughter (a college student) from serious vaccine injury.

Warning Signs

Karen is Rh-negative. In situations where the mother is Rh-negative and the father is Rh-positive, mother and baby risk being "Rh incompatible" (meaning that the child is Rh-positive), which can lead to "Rh sensitization" in the mother. For second-born and later children, this is said to increase the risk of miscarriage, stillbirth, or blood problems (called hemolytic disease) in the newborn.[235]

Since the late 1960s, doctors have urged pregnant mothers who are Rh-negative to accept "immunoprophylaxis"—injections of Rh immune globulin—to forestall these potential problems.[236] When Kevin was born in 1994, one of the Rh immune globulin products in widespread use was HypRho-D. In the mid-1970s, HypRho-D was acquired by Bayer from Cutter Laboratories, which originally developed the product. (In the 1950s, Cutter manufactured a defective polio vaccine that caused polio in at least 40,000 of the over 200,000 children who received it.[237])

Because Rh sensitization does not typically occur with a first pregnancy,[238] Karen should not have been given any shots. However, Karen's health providers gave her two thimerosal-containing HypRho-D shots, one around 20 weeks of pregnancy (most doctors wait until 26 to 28 weeks) and one right after Kevin's birth.

Thimerosal is a neurotoxin—roughly 50% ethylmercury by weight. Ethylmercury is 50 times more toxic than the mercury compound methylmercury (found in fish and produced as a by-product of amalgam fillings) and twice as persistent in the brain.[239] It was only when Kevin was around 10 years old that Karen belatedly learned that each dose of HypRho-D had contained 80 to 120 micrograms of thimerosal—the equivalent of the thimerosal content of three to five flu shots.

One of the co-creators of another Rh immune globulin called RhoGAM declared in 2018 that his invention offered "peace of mind for Rh-negative mothers," jocularly adding "How lucky can you be?"[240] After Kevin's birth, Karen was neither lucky nor peaceful, experiencing intense and debilitating mercury toxicity symptoms (not recognized as such at the time), including brain fog, inability to focus, exhaustion, and incessant ringing in the ears. She notes, "I couldn't make sense of things I was hearing." Her doctor's frivolous response was, "Of course you feel awful, you just delivered a nine-pound baby." This seemed an inadequate explanation to her, but she says, "Kevin was my first and I didn't know what to expect."

Kevin received his own thimerosal-containing shot at birth—the first of several hepatitis B vaccines. In his early months, he was fussy and didn't sleep well. By three months of age, however, Kevin had become a "very happy baby, babbling and cooing." Karen recalls being in a restaurant with him around that time and having other women exclaim over how "chatty" he was going to be as he grew older.

Karen and her husband "had no idea that anybody in the history of the world had been injured by a vaccine." Consequently, Kevin continued to receive all recommended childhood vaccines "right on schedule"—because "that's what everybody did." At the time, they did not have a home computer. Karen says, "all I knew was what the pediatrician told me."

For Kevin's birth cohort, the childhood vaccine schedule included the following in the first 15 to 18 months, with more doses administered at four to six years of age:

- Hepatitis B vaccine (three doses)
- Hib vaccine (four doses)
- DTP or DTaP (four doses)
- Oral polio vaccine (OPV) (four doses); the U.S. gradually phased out OPV beginning in 1997 and replaced it with inactivated polio shots—OPV (still widely used outside the U.S.) causes polio-like illness[241]
- MMR (one dose)[242]

CDC also added varicella (chickenpox) shots to the schedule in 1995, and Kevin was among the first children to receive one (at 12 months of age). He was also in one of the earliest cohorts to receive a four-in-one, thimerosal-containing "DTPH" shot (combining DTP and Hib).

Between three and 15 months of age, Kevin was "mellow" and had "decent eye contact," but there were "definite signs" of low muscle tone and late gross motor development. Because Kevin was a large baby (9.5 pounds at birth), Karen assumed the motor delays had to do with his size. She recalls that during trips to the grocery store when he was about a year old, she had to prop him up with stuffed animals on either side in the shopping cart. The doctors never said a word about his development.

The Tipping Point

At 15 months, Kevin received DTPH and MMR on the same day. These shots, says Karen, "were the final hit—the final blow." Kevin began to "slowly drift away. We could see that he slowly lost eye contact after that visit." However, there was no screaming or fevers; "it was way more slow and gradual," making it "hard to pinpoint."

At age two, Kevin remained verbal but "was not super conversational." However, he began reading words even before age two, a fact that "threw Karen off" in terms of recognizing his developmental challenges. She thought, "I have a little genius on my hands."

Meanwhile, Karen gave birth to her second child, another son, and this second go-round was "completely different." She felt better after the birth, without "that horrible toxic feeling or brain fog." She does not have records of what Rh-related shots she may have been given in connection with her second pregnancy, but she suspects she received less thimerosal. Kevin's brother was an easier baby, sleeping through the night; he also interacted far more with others.

The Diagnosis

When Kevin was three and his brother was one, Kevin was diagnosed with PDD-NOS. Karen and her husband had requested a full diagnostic work-up—involving assessments from a developmental pediatrician, speech and occupational therapists, and more—because they "noticed that Kevin wasn't doing the same things as his brother." When the assessment team delivered its verdict, Karen "could see from the look on their faces that it wasn't going to be a good summary." When the McDonoughs asked, "What can we do?" the team's depressing answer was, "Nothing." Karen describes it as a "very upsetting" time with a "really tough grieving process." Until Kevin's PDD-NOS diagnosis, no health provider had ever said that they thought anything was wrong.

Karen and her husband still had not connected Kevin's challenges with his vaccines. This was 1997, when information remained hard to come by, and they had only just gotten a home computer. At some point, Karen had a conversation with the mother of another special needs child, a nurse whose husband was a pharmacist. When the woman asked Karen if she thought Kevin's issues were related to his shots, Karen's first reaction was, "Of course not, because why would the pediatrician tell me to give shots to my second child if that were the case?" The other mother voiced her suspicion that, in their family's case, the shots were responsible for her child's problems, and Karen filed the conversation "in the back of her mind."

Around 2004, when Kevin was 10—with a brother who by then was eight and a new one-year-old sister—Karen was diagnosed with cancer. In the process, she experienced "horrible medical malpractice," which led her to question aspects of her situation. As another doctor helped her decipher the negligence, she recognized that what was going on was "a cover-up attempt." This prompted her to "dig in deep to figure out what happened to Kevin."

Karen says, "Once I applied myself and started doing my own research—and stopped depending on the professionals—I figured things out pretty quickly." One of the first articles Karen came across was a piece about Lyn Redwood, a nurse-practitioner who was, at the time, president of SafeMinds.[243] (From 2016 to 2021, Redwood was president of Children's Health Defense/World Mercury Project, and she now serves on CHD's Board of Directors.[244]) In 2001, Redwood and co-authors had written a landmark paper titled "Autism: a novel form of mercury poisoning," which was one of the first to link autism symptoms to mercury poisoning.[245] She notes that she found Redwood particularly credible, because Redwood "had nothing to gain and everything to lose" by speaking out on these topics.

When Karen read about Redwood's son, who had received 125 times the EPA federal safety guidelines for safe mercury exposure from his infant vaccines, resulting in a diagnosis of autism, Karen began pulling up Kevin's vaccine records and researching the shots' thimerosal content. It was at that point that the McDonoughs stopped vaccinating all three children.

As Karen studied the thimerosal issue, she learned how much thimerosal was in the Rh shots she had received during her pregnancy with Kevin. This information was "the proverbial lightbulb," allowing her to make sense out of her miserable postpartum experience. In considering lifetime mercury exposures, she also connected further dots to past dental work and mercury amalgam fillings—which pose risks of mercury accumulation in both the adult and fetal brain over time[246]—"thinking long and hard about a childhood dentist who had constantly filled her teeth."

Karen spent a lot of time educating herself. She "went to all the conferences, and jumped in really quickly in the autism movement." Because she lived in proximity to the venue housing the annual AutismOne conference—the nation's largest parent-run autism event—she attended every year.

Medical Experimentation

In 1994, Kevin's birth year, thimerosal was present in multiple childhood vaccines and notably the birth dose of hepatitis B vaccine.[247] Roughly a decade later, the CDC claimed thimerosal had been taken out of childhood vaccines. To this day, however, the neurotoxin remains in many of the flu shots given in the U.S. to babies and pregnant women, as well as in the diphtheria-tetanus vaccine. As CHD Chairman Robert F. Kennedy, Jr. puts it, "Thimerosal wasn't so much removed as it was moved around"[248] (see "Inaction and Obstruction").

For many years, inaction on thimerosal was also the order of the day for Rh immune globulin manufacturers. Although parents of injured children filed a barrage of lawsuits stemming from the product's high thimerosal content,[249] these plaintiffs experienced little success. The Johnson & Johnson subsidiary Ortho-Clinical Diagnostics, which manufactured RhoGAM, beat back lawsuits by hiring researchers to conduct gerrymandered studies exonerating thimerosal.[250]

Rh immune globulins come with other problems, too. Because they are manufactured from human blood, recipients face "a risk of contracting blood-borne pathogens."[251] In addition, although current formulations are said to no longer contain thimerosal or other preservatives, they feature the problematic surfactant polysorbate 80, a frequent vaccine ingredient associated with anaphylaxis and flagged for potential carcinogenicity.[252]

Inaction and Obstruction

In 2004, when SafeMinds' Lyn Redwood furnished testimony before a House subcommittee, she noted her shock at the government's inaction and obstruction related to thimerosal:

> *"In July 2000 when SafeMinds presented to the Government Reform Committee the paper,* Autism a Novel Form of Mercury Poisoning, *publishing the evidence pointing to the synonymous nature of the symptoms of mercury poisoning and autism spectrum disorders, we could not have imagined that in 2004 thimerosal would still be in vaccines and that the government agencies tasked with protecting the public would have failed to take aggressive action to get the mercury out and protect our nation's children. We could not have imagined that they would, instead, have focused their energies on avoiding or hiding the truth that is before them, and in doing so undercut the public's trust while continuing to put babies at risk for mercury injury."*

Source:

Testimony of Lyn Redwood, RN, MSN, President, Coalition for SafeMinds before the Subcommittee on Human Rights and Wellness Committee on Government Reform, U.S. House of Representatives, Sep. 8, 2004. Hearing: "Truth Revealed: New Scientific Discoveries Regarding Mercury in Medicine and Autism." https://www.safeminds.org/wp-content/uploads/2014/02/redwoodsafemindssept8testimonyfullfinal.pdf

Day-to-Day Health Impacts

Through third grade, Kevin attended a therapeutic day school in Chicago that was part of the public school system—"a pretty good place, all things considered." However, third grade was as far as that school went. Had the family stayed put, his next step would have been to transition to a much larger school where it seemed unlikely that Kevin's individualized education program (IEP) would be honored. To ensure adequate educational support, the McDonoughs moved to the suburbs, buying a house directly around the corner from Kevin's school so that he would not have to take a school bus.

The support of one-on-one paraprofessionals and Kevin's "savant-like qualities" allowed him to do well in school. By high school, he was winning the Geography Bee. An "outstanding" paraprofessional in high school also came to their home after school and would also take Kevin on social outings. Kevin was able to complete his academics on time and graduated with his class. After graduating high school, he stayed in a life skills program through age 21. (Young people on the autism spectrum are generally eligible to receive this type of transitional service via the school system until they turn 22.)

During his school years, functional medicine treatments helped Kevin "do really well" for a time, allowing his parents to "see who he was supposed to be." Karen recalls teachers reporting back to her, "I'm seeing a miracle—he's engaging and raising his hand." Sadly, the improvements did not last, and at a certain point, the McDonoughs watched their son "slowly slipping away again."

Karen says that Kevin "has a lot of knowledge, but you need good neurological function to use that knowledge." She describes these processing challenges as "output dysfunction."

Financial Impacts

Although Karen suffered considerably as a result of her medical malpractice experience, the silver lining was that she ended up winning a malpractice lawsuit. The proceeds provided the family with additional resources that made it possible for Karen not to work for 10 years, helped pay for out-of-pocket therapies and programs, and allowed them to purchase an income-generating property that continues to help cover expenses.

Unlike some families, the McDonoughs consider themselves fortunate to have had decent health insurance. At the time of Kevin's PDD-NOS diagnosis, the insurer was a health maintenance organization (HMO). Immediately after the diagnosis, Karen went to work and found the top pediatric neurologist in the area—who happened to be in-network with their HMO—and scheduled an appointment. After they had waited four months for the appointment, the neurologist's staff called to cancel the appointment at 4:00 PM the day before, "because they didn't want to deal with the HMO" and related paperwork hassles.

After that, the McDonoughs switched to a "very expensive" preferred provider organization (PPO), a type of insurance plan that costs more—with higher premiums and copays—but offers more flexibility and the ability to see any provider in or out of network, without a referral. On some occasions, they found that Kevin's functional medicine providers were willing to accept the PPO insurance. The family has maintained the PPO plan to this day and was able to keep Kevin on as a disabled dependent after they petitioned Mr. McDonough's employer to do so. At this juncture, however, Karen says she is ready to switch back to a basic plan that covers just major medical and hospitalization, saying she and her husband are "sick of paying for bad medicine and arrogant, abusive providers."

When he was 21, Kevin developed a brand-new symptom—"horrible," intractable epilepsy not controllable by medication. By the time of the seizures' onset, Karen had been back at work for some years, including working for the National Autism Association and serving for three years as executive director of the Autism Society of Illinois. However, the severity of Kevin's seizures prompted her to quit working again to help her son. (In 2021, Karen returned to work part-time at CHD.)

Social Impacts

In 1997, when Kevin received his PDD-NOS diagnosis, "there was no social support at all." On top of dealing with her own sorrow, Karen endured blame from relatives who didn't know how to be supportive or even thought she "must have done something to cause it." She comments, "It was devastating on multiple levels."

Later, Karen concluded that many people "don't use logic" when thinking or talking about vaccines. She also notes that most pediatricians "are so brainwashed by the AAP that there is no possibility of discussion about treatments for vaccine injury."

Karen recalls a time when she went to a rally in Washington, DC protesting the medical trade group the American Academy of Pediatrics

(AAP), one of the United States' most notorious vaccine industry front groups.[253] AAP receives funding from all four manufacturers of childhood vaccines in the U.S. and gets significant funding from CDC, over a third of which is explicitly vaccine-related. The AAP's journal *Pediatrics* routinely publishes uncritical and substandard vaccine research by authors with blatant conflicts of interest.

As it happens, AAP is headquartered in Illinois near Karen. A woman who was an AAP spokesperson had a daughter in Kevin's class; when the woman learned that Karen had attended the rally, she tried to convince Karen that AAP "cared about children." When Karen responded that if AAP cared about children, "they would help children by getting the mercury out of vaccines," the woman's position was that AAP would "never make a public statement about autism and vaccines because if we do, parents would never vaccinate again." While saying she was "sorry about what happened to Kevin," she insisted the vaccine program was "saving lives" and that "children would die" without vaccines.

Kevin Today

After aging out of the life skills program, Kevin was able to attend a nonprofit transitional program for one year, until about age 23. Next was the prospect of a vocational program through the state's Division of Rehabilitation Services; however, the program's nearly unsurmountable bureaucratic hurdles ended up causing the McDonoughs to "give up on it."

Kevin's seizures, called partial complex seizures, generally last around 30 seconds and require that Kevin sit down. Sometimes, he has several in a row. For a couple of years, Kevin and Karen made the rounds between functional medicine appointments in a neighboring state and the seizure clinic at the University of Chicago. Unfortunately, there are few options for intractable seizures. One option—a neuromodulation-based approach called a vagus nerve stimulator that Karen describes as being "like a pacemaker for the brain"—sometimes reduces seizure frequency but, as an implant, is "invasive and permanent."[254]

Over the years, Kevin's seizures have deprived him of being able to ride his bike to the local grocery store—one seizure he experienced while bike-riding resulted in surgery. More recently, another seizure-induced fall left him with a lacerated hand from a broken mug. The family went to the hospital where, ironically, a nurse tried to bully them into giving Kevin a tetanus shot, even calling security to try and kick Karen out. Fortunately, the nurse's manager was a friend of Karen's and was able to intervene.

For a while, a former school assistant worked with Kevin in a skills program covered by insurance, through which Kevin volunteered at the local library helping to reshelve books. (Kevin was very good at this and still knows all the categories for the Dewey Decimal System.) Nowadays, Kevin goes to work with his father, a firefighter, twice a week and is home the other days. His father is coming up on retirement, however, so the family is exploring business ideas that Kevin could help with.

One of Kevin's biggest challenges is loneliness and social anxiety—socializing is difficult. Though quite empathetic, he can't necessarily communicate it. Nevertheless, Karen says, he is "the first one to give someone a hug when they're upset."

Jackson's Story

[*As told by Jackson's mother, Laura Bono, on April 14, 2022. Laura is CHD's Executive Director.*]

The Overview

Jackson Bono was born with 9 and 10 Apgar scores in 1989. This was the "change year" when American children's health started going downhill—at the same time that school vaccination requirements started going up. Jackson was a bright-eyed, social baby and toddler, a "prankster" who enjoyed teasing his twin sisters (two years older than him) by running away with their blankets, and a child so perceptive and alert that he could tell the difference between a regular airplane and a Concorde. He met all his developmental milestones until 16 months of age, when a heavy load of thimerosal- and aluminum-containing shots consolidated his descent into severe autism. Jackson, now 33, lives at home with his parents and works part-time folding laundry at a local company that primarily serves a local university—but only when he has a helper or relative to take him to his job and make sure he does not wander off. The Bonos have never stopped looking for ways to help their son attain a higher level of health.

Warning Signs

Jackson's mother Laura recalls that as a child, when she got a smallpox shot to go to first grade, another little girl in her community died immediately after getting the same shot. A relative of Laura's, a health provider, administered the fatal shot and was quite shocked by the outcome. Still, the prevailing attitude was, "these things happen but it's one in a million." After that childhood incident, Laura barely ever heard mentioned the possibility of adverse events. When she had her own children, it was before the advent of the Internet, and information about vaccine risks was largely unavailable.

In the late 1980s and early 1990s, vaccination was still largely unquestioned; the general sentiment was, "it's just what people do." When Jackson got his first shots at two months of age, he was the unfortunate recipient of two vaccines with side effect profiles already known in some corners to be disastrous—OPV and the highly reactive diphtheria, whole-cell pertussis, and tetanus (DPT) shot (a combination vaccine whose "D," "P," and "T" components had started to be bundled together in the late 1940s).[255] In the 1970s and 1980s, problems with the DPT began escalating; parental reports and other data sources indicated that a large proportion of infants who received the shots were reacting to the "P" component with serious symptoms like high fever, febrile seizures, persistent crying, whole limb swelling, a "shock-like state," and brain damage (encephalopathy)[256]—reactions so severe that, for a time, countries like Japan and Sweden halted or suspended their pertussis vaccination programs.

After his two-month DPT shot, Jackson experienced some of these telltale "red flag" symptoms: high fever and persistent crying. For children such as Jackson, doctors at the time generally recommended switching to diphtheria-tetanus (DT) shots to avoid the risky "P" component—but they did not tell parents to forego the vaccine altogether. What they also did not tell the Bonos was that the DT shots contained high levels of cumulatively neurotoxic aluminum adjuvant,[257] setting Jackson up for more problems down the road. Jackson's next two rounds of vaccination at four and six months included two DT shots as well as two more doses of OPV.

The Tipping Point

At 16 months of age, Jackson received a "catch-up" Hib vaccine (containing 25 micrograms of mercury) and, two weeks later, another round of DT (with an additional 25 micrograms of mercury plus aluminum), along with another OPV and the MMR. Ten years later, in a secret gathering known as the Simpsonwood meeting convened by CDC, a statistician for the National Immunization Program commented that "an unusually high percentage" of children who, like Jackson, received catch-up Hib shots were what he euphemistically referred to as "outcomes."[258] The ostensible purpose of Simpsonwood was to discuss "theoretical" concerns about thimerosal-containing vaccines, but despite a safety signal that wouldn't "go away," all but one of the 50-plus members of the cabal agreed, by the end of the two days, to rate the association between thimerosal and neurodevelopmental disorders as merely "weak."[259]

By 16 months of age, the Bonos later calculated, Jackson's vaccines had delivered a total amount of mercury that was 139 times the EPA's "safe" exposure level for an adult, along with considerable aluminum. (Many scientists criticize the EPA's notion that there is any "safe" level of mercury at all.[260]) Both thimerosal and aluminum adjuvants are recognized for their "bio-active" ability to disrupt the neuroendocrine and immune systems, but even as late as 2020, their combined neurotoxic effects in infants remained an ignored area of research.[261]

The Diagnosis

From 16 months on, Jackson deteriorated, taking about six months to hit "rock bottom." As Laura describes it, "That huge insult took time to play itself out." By 20 months of age, a prestigious teaching hospital had diagnosed him with pervasive development disorder (PDD), which had just been added to the *Diagnostic and Statistical Manual of Mental Disorders* (DSM). Back then, information was beginning to trickle out about autism and aphasia (the inability to comprehend or formulate language because of damage to the language areas of the brain), but there was virtually no information about PDD.

At the initial time of his PDD diagnosis, Jackson had no issues with sleeping or eating, but in another two months, "the sleeping child was replaced by one who would awake startled after a few hours and not go back to sleep all night, and the one who had a healthy appetite was replaced by one who became picky and whittled his foods down to one or two." Jackson became "a shell of his old self." As his horrified parents later stated, "The happy child was replaced by an unhappy one. The child who at one time didn't miss a thing and was the life of the room became distant and preferred to be alone. The one who was developmentally ahead of the crowd began making strange noises and exhibiting odd behaviors."

As the Bonos worked with Jackson in various ways in those early months, his eyes told them that "he was scared of what was happening to him." They would show him a spoon and say, "put the spoon on the table." Jackson went from initially understanding the question and giving it a try (but perhaps putting the spoon on the floor because he no longer knew what "table" referred to) to complete aphasia—with "expressive and receptive language skills gone." Similarly, when they showed Jackson figure-ground pictures and asked him to find the figure within, Jackson went from searching for and failing to find it to no longer even grasping the request. Laura describes the sadness of observing her son's growing confusion and his heartbreaking looks that conveyed, "This used to mean something to me."

At the time, the family lived in Virginia, in the Washington, DC, suburbs. In that state, children could start school at age two, so—because school was the route to accessing services that Jackson desperately needed—he started school on the day that he turned two. "That was a sad day," Laura says. In school, his diagnosis shifted from PDD to autism, then still wrongly viewed by most doctors as strictly psychiatric rather than neurological.

One of Jackson's psychiatrists was more enlightened, however, and recognized Jackson's condition as neurodevelopmental. His diagnosis for Jackson was "static encephalopathy," a biologically-based non-progressive disorder of the central nervous system that, among other features, "derail[s] a child's overall relating and communicating."[262] This was the only physician ever to acknowledge that Jackson's problems were vaccine-related. Jackson's other doctors would just "deal with the diagnoses in front of them" and say, "Jackson has immune issues" or "Jackson had an allergic reaction to mercury."

Over time, Jackson accumulated an astounding array of diagnoses in addition to PDD, autism, and static encephalopathy:

- Allergies to foods, inhalants, and chemicals
- Apraxia ("loss of the ability to execute or carry out skilled movements and gestures, despite having the desire and physical ability to perform them"[263])
- Attention-deficit/hyperactivity disorder
- Auditory processing dysfunction
- Colitis
- Dyspraxia ("a neurological disorder that affects the planning and coordination of fine and gross motor skills"[264])
- High measles titer
- Immune system dysfunction
- Landau-Kleffner syndrome (a "rare neurological syndrome characterized by the sudden or gradual development of aphasia. . . and recurrent seizures"[265]), diagnosed at age 7
- Leaky gut syndrome
- Mercury poisoning and heavy metal toxicity
- Nodular lymphoid hyperplasia (a gastrointestinal condition)
- Pervasive development disorder
- Sensory integration disorder
- Speech, language, and social delay
- Thyroid insufficiency
- Tinnitus

- Tourette syndrome
- Yeast and viral overload

Medical Experimentation

Both the DPT and OPV vaccines attracted so much negative attention that the U.S. eventually was forced to take them off the market, replacing the DPT with the seemingly less reactive DTaP shot (by 1997), and swapping out OPV for an inactivated polio vaccine (IPV) by the year 2000 (see, however, "DPT and OPV Still Widely Used").

Soon, pediatricians also began offering combination vaccines containing both DTaP and IPV as well as additional components such as Hib or hepatitis B. Analyses of VAERS data subsequently have revealed that DTaP- and IPV-containing vaccines, too, are fraught with problems.[266]

Manufacturers justify the ongoing inclusion of neurotoxic aluminum in childhood vaccines, despite abundant evidence of harm,[267] by saying they need it for "immunopotentiation" purposes—that is, to kick-start the immune response.[268] They leave unmentioned what Dr. Christopher Exley, the world's leading authority on aluminum, has described as aluminum-based adjuvants' "migratory capabilities," including migration from the injection site to the brain.[269] A seminal 2018 study by Exley and colleagues reported finding consistently high aluminum levels in the brains of deceased individuals who had been diagnosed with autism (most of whom had died in their teens or twenties)[270]—levels representing "some of the highest values for brain aluminum content ever measured in healthy or diseased tissues."[271]

Day-to-Day Health Impacts

Following his PDD and autism diagnoses, life with Jackson became a "merry-go-round going from doctor to doctor." Laura was putting 500 miles a week on her car going around the Beltway to speech therapy, occupational therapy, doctors' appointments, and so on—"every day, there was something for Jackson."

While still living in the DC suburbs, the Bonos periodically drove to a southern state to visit relatives, and they noticed that every time they

DPT and OPV Still Widely Used

To this day, both the whole-cell diphtheria-pertussis-tetanus (DPT) and oral poliovirus (OPV) vaccines remain in use in many countries. Discussing the WHO's ongoing endorsement of DPT, one author states, "In spite of its reactogenicity and the occurrence of adverse effects after immunization, this type of vaccine [DPT] continues to be produced throughout the world due to its effectiveness in the prevention of pertussis. That is why WHO still recommends its inclusion in national immunization programs independently of the emergence of acellular vaccines which are less reactogenic."

Source:

Ponce-de-Leon S. Challenges in vaccine access and delivery for developing nations. *Infectious Disease Advisor*, 2017. https://www.infectiousdiseaseadvisor.com/home/decision-support-in-medicine/hospital-infection-control/challenges-in-vaccine-access-and-delivery-for-developing-nations/

did so, Jackson would "sort of wake up and begin to say words from his environment," for example "smoke" when passing through a big city, "cow" when passing by fields with cattle, and so on. "It was mind-blowing, and it happened over and over." At the relatives' house, he would also pay more attention and say words that he hadn't said in a long time. Yet when the family returned to northern Virginia, over about a four-day period he would sink back into his "fog."

The immunologist they were seeing at the time opined that Jackson's immune system "had become allergic to his world, and when he got out of his world, he got better." The doctor also speculated that DC air pollution might be a significant culprit. As a result, the family decided to move to Durham, North Carolina. At a time when it was difficult for someone in the insurance industry (Jackson's dad) to pick up and move his business, and almost unheard of to work from home (as Laura began to do), they were able to make the move around Jackson's sixth birthday. Laura says, "It is one of the best things we ever did for him."

At the time, Jackson was not yet potty-trained and was having diarrhea multiple times a day. The sudden diarrhea would run down his legs and literally blister them. The Bonos made sure to always have water on hand to try to prevent the blistering. Around the time of their move, they took Jackson to see a doctor out of state who diagnosed issues with yeast and viruses. After Jackson began taking antifungals and antivirals, "they saw amazing things and Jackson got better and better." When he started on an antiviral, he potty trained in three days and started having formed stool. They also put Jackson on a gluten-free diet. All of these interventions "came together for him to feel better," with less brain fog.

Around this time, the Bonos were also getting on email lists and learning about other resources. They soon consulted with a California pediatrician who agreed that ASD was a neuro-immune disorder, not a psychiatric disorder. He gave the Bonos a long list of tests and asked them to get their local doctor (at Duke Medical Center) to run the blood work.

After Duke started getting the first test results back from the list of immune system panels, blood work, and other panels, Duke called the Bonos and said they wanted to rerun the tests, even saying they would pay for them. Jackson returned for more blood draws and Duke sent the tests off to the Mayo Clinic. The results came back the same.

Jackson's doctor at Duke said, "Your son has complete immune system dysfunction—some of the worst we've ever seen." These oddities included high eosinophil counts, a high measles titer (122 times over normal) at SSPE levels (subacute sclerosing panencephalitis is a progressive neurological disorder thought to be caused by "defective measles virus"), a low erythrocyte sedimentation rate (a potential indicator of blood cell abnormalities), zero antibodies to yeast, and more. From that point on, the doctor and his head nurse (someone Laura says she will "always be grateful for") were willing to request any test that the Bonos asked for. The doctor also wrote letters stating that Jackson (and his sisters) should not get any more vaccines.

It was not until age 11 that Jackson was diagnosed with heavy metal (mercury/thimerosal) poisoning. The diagnosis helped make sense of many other symptoms witnessed over the preceding years. The doctor who diagnosed the poisoning, herself a physician-mother of an autistic child, was one of the first to treat autistic children using chelation. Jackson responded positively to chelation, which also allowed his immune system to get better, including his "overwhelming" allergies to chemicals, inhalants, and foods. The Bonos later credited chelation with helping Jackson "make greater gains from age 11 to 20 than he did from 2 years to age 10"—with better eye contact, behaviors, talking, and cognitive, social, and self-help skills.

"Blown away" by Jackson's immune system issues, Jackson's doctor at Duke successfully lobbied for him to get monthly intravenous immunoglobulin (IVIG) treatments. Shockingly, the Bonos learned that all but one of the IVIG products on the market at the time contained thimerosal, and they were probably making many children with epilepsy or autism worse, not better.[272] The one IVIG product that did not contain thimerosal was more expensive, but Jackson's doctor made sure he got that one.

During Jackson's school years, the Bonos "fought for everything," developing a reputation as parents who would go to bat for their son. He had an IEP from day one. Fortunately, Jackson was easy-going and did not have some of the behavioral issues often seen in children with autism.

In general, school proved to be a good setting for him. In second grade, however, Jackson started coming home and hitting himself in the head. The Bonos knew that someone must be hitting him at school (or that he was witnessing another child being hit), telling the principal and teacher "this is a child who doesn't learn unless it is shown to him," and he had never done it before. Although they brought the matter up repeatedly with the teacher and principal, the school denied that anything of the kind was happening.

On the last day of school, the class went on a field trip to some public gardens, accompanied by two aides and the teacher. As it happened, a local doctor who had an autistic child himself was sitting in the gardens having lunch, and noticed the class because of his personal interest in autism. He observed Jackson take his snack (popcorn) out of his backpack and then was shocked to witness one of the aides grab the popcorn, smack Jackson on top of the head (causing Jackson to cry), and then eat the popcorn herself. The Bonos later learned that this had been happening all year. After the good samaritan doctor reported what he had seen, the school principal called the Bonos "with his tail between his legs." The Bonos were able to obtain a written statement from the witnessing doctor and filed a lawsuit, not asking for monetary damages but reserving the right to sue the Durham public schools for abuse and assault for seven years (the statute of limitations). After that, Jackson's IEP and education in general got special attention from school management, including providing an aide. Sadly, it took about three years for him to stop hitting himself.

Financial Impacts

Jackson was born in 1989, and it wasn't until the early 2000s that health insurers would even begin to cover anything autism-related (before then deeming autism purely a "mental disorder"). For the Bonos, it was "money, money, money—mostly out of pocket." Speech therapy, occupational therapy, play therapy—very little, if any, was covered.

When Jackson received his static encephalopathy diagnosis, it helped get some things paid for, including some speech and occupational therapy, because it was a "medical diagnosis." Around age 12, however, insurance stopped paying for occupational therapy and never would again. Around that time, the occupational therapy venues also began turning Jackson away, saying he was "too big for the equipment."

When Jackson was in kindergarten, his teacher helped Jackson at home for two hours every day after school. He recognized that Jackson was smart but needed more help, and he himself needed the extra income. At $50 per school day, that added up to another $1,000 or so per month. However, having the in-home help allowed Jackson to learn while making it possible for Laura to get some of her work done.

The Bonos and other autism families learned various ins and outs of the system. For example, if Jackson's conventional doctor ordered a particular test, insurance would pay for it, but if an "alternative" doctor (one who perhaps acknowledged the vaccine damage) ordered the same test, "nothing doing." Having a conventional doctor who would "go to bat" for Jackson made a world of difference; when he began doing IVIG, the Bonos paid about $2,000 per month though IVIG cost $15,000 per month.

Especially in the early years after Jackson's injury, the Bonos "desperately needed" Laura's income—she couldn't quit working. She counts herself lucky that she had "the most amazing boss ever," an employer who permitted working from home and a flexible work schedule because he "knew she would get the job done." Laura routinely worked late into the evenings and on weekends.

The Bonos did not originally know about the NVICP. When they learned about it, Jackson was out of the short three-year statute of limitations. Still, they filed a claim, and when the program said it would lump autism claims into the OAP, they even helped other families file NVICP claims. Laura recounts:

> "*We knew we would be dismissed from the OAP because of the three-year statute of limitations, and in fact, we were in the first group to be kicked out. By the time we put the claim in, Jackson was 13 or 14 and it had been well over a decade. But we did it anyway to get his injuries into the official record. We thought, 'We are going to put this on the record. The government is going to know that we believe Jackson was vaccine-injured.'*"

Laura notes that the NVICP used many shenanigans to reject people's claims. For example, if the petitioner's injury exceeded the three-year statute of limitations by one day, they would be kicked out. The program would also "play games" with children's medical records: "You took Suzy in for eye fluttering when she was 14 months old, but she didn't get an autism diagnosis until she was four. The three-year statute of limitations dates from the first symptoms, not from the diagnosis; therefore, you're excluded."

Insultingly, medical figures who persist in denying that vaccines (and ingredients like thimerosal) have the potential to cause harm sometimes try to deflect blame for their products' inherent risks on parental greed,[273] ludicrously suggesting that families like the Bonos—who are prevented by the 1986 Act from suing the manufacturers of Jackson's vaccines and who were denied any government compensation—are like "sharks with blood in the water," engaged in a "feeding frenzy" driven by money. The Bonos, who by the time Jackson was 21 estimated having spent roughly $50,000 of after-tax income per year out of pocket—almost $1 million over two decades—have many times explained that "they do not want blood," but "want families like theirs to be heard for Jackson's sake, and others like him."

Over the years, the Bonos spoke frankly with their twin daughters about finances. They explained, "We are giving the most time and money to the one most in need. Please don't feel resentful or slighted—resentment will eat you alive. Instead, be thankful that you don't need the help, and be grateful that you are smart and can get scholarships" (which, indeed, they did). Jackson's sisters will be his legal guardians when their parents are gone; "they know it and are up for the task." The Bonos have planned ahead and already own a house that at some point will be "Jackson's house."

Social Impacts

Laura describes severe autism as extremely isolating. For the Bonos, there were no "fun family outings." If they did go out, they would have to closely supervise Jackson—or chase him, doing what they called "zone defense"—making sure he did not eat something he was allergic to or worse, sand or worms, any of which would give him "diarrhea all night." Going out and seeing "other kids doing what Jackson should have been doing" was also depressing. Even among other autistic children, it seemed like her son was always the most severely affected in the group. At home, they needed an alarm system to make sure Jackson did not slip out without their knowledge.

As a family, they couldn't even do simple things like throwing a ball, because Jackson had no concept of "back and forth." If he threw the ball once, he thought "that was that." Because he liked the ABCs, they found that they could throw the ball saying "A," "B," and so forth, but when they got to "Z," that was it.

With some of Laura's oldest friends, there was a "dead zone" for many years. At the time, people did not understand what the Bonos were going through. She says that now, "people are finally realizing that there is such a thing as vaccine injury; they are hearing more about it and thinking, 'maybe Laura wasn't such a crazy person after all.'" She is just now circling back to some of those friends.

Eventually, the Bonos developed friendships with people who "got it" and with whom they could spend some enjoyable time. On one occasion, when a neurologist friend and spouse came over for dinner, a nude, 12-year-old Jackson walked through the living room. The couple was completely unfazed. Having an autistic son themselves, they lived with these complexities of life. Everyone had a good laugh after Jackson was required to get dressed.

Jackson's sisters were and are "his greatest champions." Laura cites them as models of unconditional love: "He couldn't give anything, but they loved him anyway—protected him, included him, and always would work with him." When they were little, they would ask, "Why won't Jackson talk to us? He doesn't even want to play with us." When the girls were four or five, the Bonos decided to tell them about the shots, letting them know that Jackson had been injured and that "we all need to help him learn." The girls started crying from relief that it was not anything *they* did. Laura says sadly, "They were taking the guilt on." During the girls' younger years, the Bonos had a therapist check in with them once a year to make sure they were okay, and "thankfully, they always were."

Once the Bonos learned the truth about vaccines and mercury (among other vaccine dangers), they became autism activists. Laura became involved with Unlocking Autism (founded in 1999 to "raise awareness about autism spectrum disorders and bring our community's issues from individual homes to the forefront of national dialogue"),[274] and Laura became its representative in North Carolina. Later, feeling the lack of an advocacy organization linking autism and vaccines, both Bonos served as founding board members of the National Autism Association (NAA).[275]

(Laura, along with other founding NAA board members Lyn Redwood and Rita Shreffler, were among those who went on to help launch World Mercury Project, which is now CHD.) NAA began doing joint press releases with SafeMinds,[276] founded in 2000 with a focus on mercury, and they began building a coalition.

About activism, Laura says: "Depression and anger are 'inside out' with each other. The activism gives me a place to put the depression." She thinks, "I'm going to get those guys today!" At the same time, she admits that "Up until the Simpsonwood documents came out, I was under the mistaken impression that if we put on our pearls and pantyhose and went to meetings at the CDC, FDA, NIH, and NIEHS, we could tell them what was happening with children and vaccines, and then they would stop. If we could tell them they're poisoning our kids and need to stop, they would stop. Simpsonwood showed that they knew—and weren't going to stop." She realized then what they were up against—a trillion-dollar industry.

Jackson Today

Jackson is a "happy, kind, and emotionally attached" 33-year-old, but "the Jackson that could have been—the brother and son his family could have had for all these years"—was lost forever after his 16-month shots. He has no friends other than his family and no one to spend time with him unless they are paid.

With the retirement of his long-time Duke doctor, Jackson no longer has a physician who is willing to go to bat for his health issues. Jackson still has digestive problems, including irritable bowel disease (IBD) and colitis, and has to take a gut anti-inflammatory to avoid motility issues, but his current doctors will not renew his prescriptions unless he gets an annual colonoscopy (a procedure that comes with its own set of risks).[277] Although Jackson also benefits from periodic antifungals, again, his doctors are less ready to prescribe them.

Jackson is overweight and often asks for more food. Since he was a young child, anything that he might eat has to be kept in the basement refrigerator or under lock and key in the kitchen; otherwise, he will eat it, including plain flour or sugar. Laura believes his incessant hunger is tied into his inability to produce the pancreatic stimulation hormone secretin, and as a result, CCK (cholecystokinin). CCK, the GI hormone responsible for stimulating digestion of fat and protein, is usually released 20 minutes after a meal begins and reduces appetite.

Currently, Jackson gets two forms of government assistance: Social Security Disability and a Medicaid waiver (the Community Alternatives Program for Disabled Adults or CAPS). Jackson has a job folding laundry, but he can't go to work without a helper. In theory, CAPS is there to cover the helpers, but since COVID, the helpers have evaporated (either because they were unwilling to come to the house or because they got scooped up for better jobs). He needs a helper to keep him on task. He is bright enough to do all the washing, drying, and folding himself, but he needs someone to keep him focused; otherwise, he might disappear or go looking for food. During the two years of COVID, Jackson only went to work when either his dad or his aunt could take him, usually just for two to three hours at a time. Otherwise, he was stuck at home. And, boredom has its price. Recently, Jackson discovered the gas pilot light in the fireplace and throws paper in "to see what happens." His family still needs to keep a close watch over him, including keeping the alarm system activated 24/7 when he is at home.

In 2010, the Bonos wrote the following:

> *"What Jackson, and his family, will NEVER get back is his childhood. Although we are so thankful to have experienced Jackson in any form, we have truly missed out on the most social, smart kid we could have ever known if he had continued to follow all the signs he was showing at the young age of 16 months old. We were robbed. We feel violated every day of our lives. Maybe it is just the sad musings of a couple of parents, but we really believe thimerosal and the neurodevelopmental/immune system dysfunction it brings with it, have taken society's best and brightest from us.*

CHAPTER FOUR

Childhood Vaccine Injuries: The Children Are Not Alright

NORMALIZING ILL HEALTH

As discussed in Chapter One, the proportion of American children beset by one or more chronic conditions has been increasing at an alarming rate—with potentially life-threatening conditions like type 1 diabetes and cancer affecting children at younger and younger ages.[278,279] Large percentages of children also suffer from less dramatic but still concerning problems like eczema (15% of American children)[280] and constipation (responsible for one in 20 pediatric doctor's visits)[281]—afflictions that not only have become commonplace but which medicine has normalized as "part of childhood."

Even in the absence of formal diagnoses, many children are simply unwell, lacking the stamina, resilience, and cheerfulness that should be all children's birthright. The following story illustrates one child's "what's wrong with this picture" trajectory—describing a peculiar series of problems arising in tandem with vaccination, even with vaccines administered on a modified and delayed schedule. Fortunately, the parents were able set their daughter on a different path early enough to make a significant difference.

Chronic Illness Is . . . Stressful

An often-quoted child health study defines chronic health conditions among children and youth as "any physical, emotional, or mental condition that prevent[s] [a child] from attending school regularly, doing regular school work, or doing usual childhood activities or that require[s] frequent attention or treatment from a doctor or other health professional, regular use of any medication, or use of special equipment."

As this definition suggests, chronic illness equates to chronic stress, both psychological and physical, including multifaceted stress arising from changes to daily routines, frequent interactions with the health care system, treatment side effects, and uncertainty about long-term prospects. Significantly, the stress affects not just the child but the entire family.

Sources:

Compas BE, Jaser SS, Dunn MJ, Rodriguez EM. Coping with chronic illness in childhood and adolescence. *Annu Rev Clin Psychol.* 2012;8:455-480.

Van Cleave J, Gortmaker SL, Perrin JM. Dynamics of obesity and chronic health conditions among children and youth. *JAMA.* 2010;303(7):623-630.

Giuliana's Story

[*As told by Giuliana's mother, Melissa Bordes, on May 3, 2022. Melissa works with CHD's Chapter Program (including both national and international chapters).*]

The Overview

Giuliana Bordes was born in 2012 in New York State. From day one, she displayed a range of alarming reactions to childhood vaccines. Although her mother Melissa did not immediately connect her daughter's problems to vaccination, she repeatedly questioned medical providers as to whether such reactions were "normal." The clear response was, "don't question—the doctors know more than you do." Although Giuliana received around 20 different doses by age three, Melissa credits the family's decision to give the vaccines one at a time with lessening the assault on Giuliana's developing immune system. Around age three, a dose of MMR caused Giuliana to experience excruciating two-day migraines. At this point, the family stopped vaccinating and began working with a holistic pediatrician—two steps that were critical to helping Giuliana recover and become the healthy, straight-A student that she is today. The family left New York and moved to Florida in 2021.

Warning Signs

As a new mother, Melissa "knew nothing" about vaccination issues. She says, "I did it because doctors told me to." In her extended family, she was the trailblazer—the first to have a child and, therefore, the first to venture into decisions about childhood vaccines.

On the day of Giuliana's birth, nurses took the baby away to administer hepatitis B and vitamin K shots. Right after they brought Giuliana back, Giuliana suddenly turned blue and stopped breathing. Alarmed, Melissa asked, "Is this *normal*?" The nurse slapped Giuliana to get her to breathe again and merely said, "yes," even though symptoms such as bluish skin (an indication of low oxygen saturation) and respiratory distress are known signs of a possible allergic reaction to a vaccine.[282] After being told that cessation of her baby's breathing was "normal," Melissa was terrified it would happen again and was on edge for many days.

At two months, Giuliana had her first "well-baby" appointment. Melissa remembers the appointment vividly, including her own "visceral reaction" and nausea when she saw the health workers bringing in a bunch of shots for her daughter. Giuliana received "everything they normally give" at two months: a second dose of hepatitis B and initial doses of rotavirus, DTaP, Hib, polio (IPV), and pneumococcal conjugate vaccine (PCV).[283] Melissa tried to dismiss her ongoing unease by telling herself, "maybe I'm overthinking this."

After the two-month appointment, a friend of Melissa's husband suggested not letting Giuliana get more than one shot per visit, sharing the perspective of parents "who still want to vaccinate but just not on the CDC schedule." The Bordeses heeded the friend's advice; from that point on, Melissa says, "Every time I would go in, I would give her just one," while leaving the decision of which one to administer up to the pediatrician.

Around that time, Melissa read *The Vaccine Book* by Dr. Bob Sears,[284] which included information about alternative vaccination schedules that helped to reinforce her decision. Ten years later, she wishes she had read Neil Miller's *Vaccines: Are They Really Safe and Effective?* or *Miller's Review of Critical Vaccine Studies* in addition or instead.[285,286]

At around three months, Giuliana started having "strange episodes," balling up her fists, turning bright red, not being able to move, and having her eyes fixed in one direction. According to Melissa, "It looked like she was 'stuck,' that's the only way to describe it." Shouting or snapping fingers would get no reaction from Giuliana. Each time, Melissa rushed her daughter to the doctor, only to have her concerns dismissed.

In addition to the seizures, which occurred sporadically, Giuliana was constantly sick. Melissa continued to second-guess herself: "I figured, kids get sick all time, maybe this is normal." Gradually, however, she also began noticing a pattern—every time Giuliana got a shot, she would get sick afterwards. For example, at about six months of age—one week after a "well-baby" visit involving vaccination—Giuliana ended up in the emergency room (ER) with a whole-body rash; health workers chalked it up to a "carrot allergy."

Giuliana saw a neurologist at one year, but despite the ongoing seizures, there was no diagnosis and no follow-up. As Melissa describes it, the neurologist and other doctors would say, "Giuliana looks okay to me" and treat Melissa "like she was crazy." Eventually, Melissa's personal research pointed to the possibility that her daughter might be experiencing absence seizures ("brief, sudden lapses of consciousness").[287]

Melissa kept trying to find a pediatrician who would be receptive to her questions and able to give her some answers, making the rounds of nearly a dozen pediatricians in all. In her search, she recalls interviewing one pediatrician and, without disclosing whether she planned to vaccinate or not, asking: "I'm curious, are there risks associated with these shots?" As soon as she left, the office called her and informed her that she would not be allowed in the practice. Melissa said, "You mean you're not going to see my daughter just because I *asked a question*?" Their answer: "That's right."

When Melissa took Giuliana to a pediatrician reputed to be an "amazing diagnostician," he diagnosed Giuliana with an obscure gastrointestinal disorder called Sandifer syndrome (associated with reflux)[288] and put her on the heartburn drug Zantac. The drug made no difference. On another occasion, Melissa took Giuliana to a pediatrician, concerned about her daughter's high fever (105°F); with a "blank stare" and no response to Melissa's question ("What's going on with Giuliana?"), the doctor merely sent them on to the ER. Melissa, "living in fear nonstop," says she got "non-answer after non-answer."

The Tipping Point

For the first several years, Melissa would not allow the pediatricians to give Giuliana the MMR or chickenpox vaccines. Her sense of caution about the MMR came from "random things people said in passing," and friends who said "we've heard about side effects." She also noticed more posts on social media about reactions to the MMR than for other shots.

Melissa remembers asking one particularly hostile pediatrician about the safety of the MMR, having heard by this time that it could be dangerous. His response: "I won't talk to you about that—don't listen to everything you hear. She'll be fine." Melissa had previously refused to let that pediatrician give Giuliana a flu shot; when Giuliana happened to get sick a month later, the doctor scolded Melissa, saying, "it's because you didn't give her a flu shot." He then tested Giuliana for influenza (which she did not have), shoving a swab "way up" Giuliana's nose in a punishing manner. Referring to one of the first celebrities to publicly question the safety of vaccines, he also denigrated Melissa as a "Jenny McCarthy activist."

As Giuliana approached her third birthday, a pediatrician told Melissa there was a measles outbreak in New York City and lobbied for Giuliana to get the combined measles-mumps-rubella-varicella (MMRV) vaccine. The doctor talked Melissa into a shot. Melissa declined the MMRV, however, opting for the MMR instead, all the while "feeling sick to her stomach." The next day, Giuliana had a rash from head to toe and a fever of nearly 106°F.

At this juncture, with Giuliana having received over 20 shots since birth, Melissa and her husband decided, "That's it—she's never getting another shot." None of Giuliana's mainstream pediatricians would ever link any of Giuliana's issues to her vaccines.

The Diagnosis

After the MMR, Giuliana stopped getting the strange seizures but started getting horrendous migraines that would last anywhere from 24 to 48 hours. When she had a migraine, she "wouldn't open her eyes for two days straight." During these alarming episodes, Giuliana also would wake up in the middle of the night vomiting. Melissa could tell there was "something going on in Giuliana's brain" and, knowing that headaches and vomiting can be signs of a pediatric brain tumor, was quite worried.[289] The family saw another neurologist who ordered an MRI, and thankfully, they were able to rule out a brain tumor.

The severity of Giuliana's migraines and the decision to stop vaccinating persuaded Melissa that it was time to find a doctor who could "look at this in a different way." After finding an integrative physician 45 minutes away, the family embarked on a holistic path that gradually weaned Melissa off the "fear-based life" she had been living, with a new focus on Giuliana's detoxification and healing.

Medical Experimentation

The medical community is fond of applying the word "unexplained" to medical events that it cannot readily pigeonhole into an existing diagnostic category. Thus, pediatricians call incidents of the type experienced by Giuliana after her birth dose of hepatitis B vaccine—when an infant "stops breathing, has a change in muscle tone, turns pale or blue in color, or is unresponsive"—a "brief resolved unexplained event" or BRUE.[290] The AAP coined the term BRUE in 2016 as a replacement for ALTE ("apparent life-threatening event").

AAP explains that BRUE is a "diagnosis of exclusion . . . used only when there is no explanation for the event after conducting an appropriate history and physical."[291] The group advises pediatricians to evaluate "higher-risk infants who have experienced a BRUE," specifying criteria for "higher-risk" such as premature birth or being under two months old. For such children, risks to be assessed should include child maltreatment, feeding problems, swallowing difficulties, heart arrhythmias, infections, birth defects, airway obstruction, and epilepsy.

Although AAP emphasizes the importance of understanding events preceding a BRUE, its checklist conspicuously omits any mention of possible iatrogenic causes ("iatrogenic" refers to conditions caused by a physician or health care provider)—such as, in Giuliana's case, the hepatitis B and vitamin K shots administered immediately prior to the event (see "See No Evil, Hear No Evil, Speak No Evil"). This, despite the fact that the VAERS database includes several thousand adverse event reports filed on behalf of infants under one year of age who experienced breathing problems or cyanosis (skin turning blue) following hepatitis B vaccination.[292,293] If the blasé health providers present at Giuliana's birth are any indication, such reactions are likely to be vastly underreported.

"See No Evil, Hear No Evil, Speak No Evil"

In 1999, the Subcommittee on Criminal Justice, Drug Policy, and Human Resources of the congressional Committee on Government Reform held a hearing titled, *Hepatitis B Vaccine: Helping or Hurting Public Health?*—exercising its "oversight responsibility for the Department of Health and Human Services." Introducing the hearing, Representative John Mica admitted that it was "the first oversight hearing on [the 1986 National Childhood Vaccine Injury Act] held in 10 years."

Rep. Mica began by outlining "shocking" information from a report about hepatitis B vaccination, which showed that adverse reactions to the vaccine (in children age 10 and under) "were 16 times greater than the cases of the disease," while there were four times as many child deaths related to the vaccine as cases of the disease. Among the parents who had the opportunity to speak at the hearing, one parent provided the following testimony:

> *"[My son] was born full term, normal in every way. Vaginal birth with no interventions or drugs. His Apgar scores were 9 and 10, which means that all of his reflexes were perfect and present. Before discharge, he was immunized with Recombivax HB against hepatitis B. . . . His fourth night in this world was his first at home. And about 5 hours after arriving home, he had his first seizure. Frantic calls to maternity and pediatric staff fell on deaf ears. . . . No one mentioned the vaccine. No one expressed concern that he was turning blue, that he couldn't stop screaming or that he appeared to be having tremors or full-body spasms. . . . [He] had 3 more seizures. . . in the next 8 days. . . ."*

On day 12, the family rushed to the ER after their child turned blue and lost and "did not resume" consciousness; they were sent home "with a shrug," again with no mention of a possible connection to the hepatitis B vaccine. The child's medical record went on to feature the "gross understatement that his first days of life prior to this [hospital] admission were uneventful."

Source:

Hepatitis B Vaccine: Helping or Hurting Public Health? Hearing before the Subcommittee on Criminal Justice, Drug Policy, and Human Resources of the Committee on Government Reform. House of Representatives, One Hundred Sixth Congress, First Session, May 18, 1999. Serial No. 106-97. https://www.govinfo.gov/content/pkg/CHRG-106hhrg63308/html/CHRG-106hhrg63308.htm

Day-to-Day Health Impacts

With their integrative physician's guidance, the family adopted a gluten-free, casein-free diet and focused on restoring Giuliana's gut health.

Melissa also began a gradual process of education and dot-connecting. However, when her second daughter was born in 2016, she had not entirely shed her nervousness. Their doctor now jokes with her about how he had to "talk her off the ledge" because of her ingrained habit of fear. She says, "I was so afraid of germs. That's how my mom was with me; that's all I ever knew." Through conversations about topics ranging from terrain theory to the neurotoxicity of aluminum adjuvants in vaccines, a world of new insights about health and illness opened up for Melissa,[294][295] leaving her continually wanting to know more. She attended lectures and joined a group of parents embarked on a similar learning journey. Eventually, she came to appreciate that a holistic approach encompasses different ways to strengthen health, including eating the right foods and having faith in God.

Initially, Melissa experienced intense pushback from relatives: "What do you mean you're not giving Giuliana Tylenol? What do you mean you're not giving her antibiotics?" Melissa says she "stuck to her guns," but it was "really, really hard." However, the results of the new approach gradually bore fruit. For example, Giuliana—who used to get strep throat twice a year—has not had it in the five years since she stopped taking antibiotics.

Melissa describes Giuliana and her younger daughter as her own in-house "vaxxed–unvaxxed" study. Whereas she accumulated "stacks of health paperwork" for Giuliana, she has no paperwork for her youngest, who has "never been on a pharmaceutical in her life."

Financial Impacts

Melissa says that in their family's case, the vaccine injury issues "weren't so much monetary as emotional." Initially, insurance covered the conventional care that Giuliana received (including the 20 doses of vaccine). When the family stopped going to what Melissa now refers to as "poison centers" (pediatricians) and began conferring with holistic practitioners, they began paying out of pocket. They are quite willing to pay for good-quality care, particularly since they do not need to go nearly as often as before and can often handle things at home. Melissa estimates that they spend about $1,500 per year on functional medicine and holistic and chiropractic doctors.

In 2016, Melissa became involved in advocacy and lobbying work, educating other parents as to the importance of informed consent and the risks of vaccination. When a bill was introduced to protect New York's vaccine religious exemption, she educated parents on ways to lobby legislators to support the bill. Unfortunately, in June 2019, the New York legislature took the opposite step, repealing the state's 50-year-old religious exemption to vaccination requirements for schoolchildren in public and private schools and daycare. Melissa became even more active, serving as an officer with the New York Alliance for Vaccine Rights (a nonprofit that "seeks to educate families and individuals about their vaccination rights and how to protect them through political action").[296]

In May 2022, parents who mounted a legal challenge to the New York lawmakers' removal of the religious exemption—arguing that the repeal violated sincerely held religious beliefs and denied children "the right to attend any manner of in-person schooling"—received the unwelcome news that the U.S. Supreme Court would not hear arguments and would allow lower-court decisions endorsing the repeal to stand.[297]

In October of 2019, as a result of the repeal, Giuliana was kicked out of school. Melissa says, "no one—teachers, the principal, the superintendent—batted an eyelash." Melissa recounts the "vile" and "humiliating" experience:

> *"Giuliana was no longer allowed in the building. It was the most demeaning experience, and the discrimination was on a level I had never experienced in my life. As soon as that bill passed, my children were considered second-class citizens. At a Board of Education meeting, I was ignored and treated like I had some sort of disease. One woman said my children put her children at risk, and she was happy we were no longer allowed in the school. My daughter, who was used to being in school with a teacher and with the friends she had known since she was a baby, was now in a situation that was hard for her to grasp."*

Forced into homeschooling (while still having to pay taxes for a denied public school education) was emotionally traumatic, and in addition, a "huge financial strain." Melissa was a stay-at-home mother at the time, and her husband had to work overtime to cover the new homeschooling expenses. Melissa estimates that they spent about $1200 per month to cover a two-day-a-week homeschool co-op, homeschool curricula, and other activities to address their daughters' educational needs.

In the 2020 elections, Melissa volunteered hours of her time to fundraisers and other activities in support of a local candidate strong on medical freedom issues. When "pretty much all medical-freedom-supporting candidates lost," the family decided—despite a 60% pay cut for Melissa's husband—to move to Florida. Though it was not easy, and they miss family and friends, she says "it was the best decision we ever made."

Social Impacts

Overcoming their initial resistance, Melissa's parents and in-laws have come to respect and support the family's health decisions. Melissa also appreciates her husband's willingness to work with her to "put the puzzle pieces together." Melissa enjoys solid support from her sister, who values the holistic approach the family have adopted. Her sister was living with them when Giuliana was a baby and accompanied Melissa to many of Giuliana's doctors' visits. "We both knew it wasn't normal."

As for friends, Melissa says that by the time she stopped vaccinating, she had largely "weeded out" her friendship circle. She says, "I don't have much in common with people who aren't willing to think about this topic. We're operating on a different frequency." She adds, "If you can't question what's going on right now, there's something wrong with your critical thinking."

Giuliana Today

Giuliana still has some gut issues and is affected by gluten. She also has a lingering issue with "lazy eye" (amblyopia), but otherwise, "she looks like a normal kid." At school, she is earning straight As. Giuliana knows the story of what happened to her and "remembers when she used to get sick all the time."

Melissa is thankful for Giuliana's recovery, adding, "I don't know if delaying made a difference for her, but based on what I've seen, children who receive a slew of vaccines in one day are worse off." Melissa believes that dietary changes and detoxification made a world of difference for her daughter.

Melissa regrets that no one told her about the risks. Her initial forays into activism began with posts on social media to let people know that there *are* risks. She says, "That's why I became so vocal—it's why I got involved. I have to let everyone know. God gave me a mouth, and I'm going to use it."

Melissa also notes that some people get stuck pondering the question, "Why are they doing this to our children?," adding that such people are skeptical that the attacks on children's health "could really be this big" or this intentional. She recommends that the skeptics look at "where the narrative began, and the billions of dollars behind the vaccination program. The only way to stop it is if people educate themselves, rise up, and refuse to let them experiment on our children."

CHAPTER FIVE

Childhood Vaccine Injuries: Experimental COVID Shots

A NEW TIDAL WAVE OF CHILDHOOD AND ADOLESCENT INJURIES

COVID vaccine clinical trials and the EUA rollout of experimental COVID injections have triggered an unprecedented wave of serious vaccine injuries among young and old alike.[298] Literally adding insult to injury, the individuals suffering the damage—to their bewilderment and shock—have found themselves hitting up against the same gamut of official responses long endured by autism families, ranging from denial to indifference to "it's all in your mind" brush-offs to outright hostility.[299] Sadly, the painful lessons learned by autism families are more resonant and relevant than ever.

Since the vaccines' approval for children, young people have experienced harm on a staggering scale. As of May 2022, VAERS had collected nearly 11,000 adverse event reports for American 5- to 11-year-olds (over six months) and almost 32,000 reports for 12- to 17-year-olds.[300] Given what we know about VAERS underreporting, the actual number of injured children and youth is likely far higher.

Despite COVID risks for children that are practically nil,[301] manufacturers and the FDA have rolled out the pediatric clinical trials and EUAs with unseemly zeal, steamrolling parents with unsubstantiated assertions about "benefits that outweigh risks." Consider the rushed timeline for the Pfizer injections, which were, until June 2022, the only ones authorized in the U.S. for those under age 18:

- **December 11, 2020:** FDA grants EUA status to Pfizer-BioNTech COVID vaccine for ages 16 and up.[302]
- **January 2021:** Pfizer completes enrollment of Phase 3 clinical trial participants, including 2,260 teens aged 12 to 15.[303]
- **March 25, 2021:** Pfizer launches pediatric clinical trial with 144 children ages six months to 11 years.[304]
- **May 10, 2021:** FDA extends Pfizer's EUA status to 12- to 15-year-olds.[305]
- **October 29, 2021:** FDA extends Pfizer's EUA status to 5- to 11-year-olds.[306]
- **December 8, 2021:** FDA authorizes an EUA booster dose for youth 16 to 17 years of age.[307]
- **January 3, 2022:** FDA authorizes an EUA booster dose for 12- to 15-year-olds.[308]
- **March 28, 2022:** FDA authorizes a second EUA booster dose for immunocompromised youth ages 12 and older.[309]
- **May 17, 2022:** FDA authorizes an EUA booster for 5- to 11-year-olds (based on study results from a mere 67 children).[310]
- **June 17, 2022:** FDA authorizes Pfizer (and Moderna) vaccines for children down to 6 months of age.[311]

Throughout this period, conscientious doctors and scientists have been pushing back against the folly—and criminality—of giving the experimental COVID jabs to children and youth.[312,313,314] As the following two stories sadly and forcefully illustrate, their warnings were only too prescient; despite the incessant propaganda, the benefits most certainly do *not* outweigh the risks.

Maddie's Story

[*As told by Maddie's mother, Stephanie de Garay, on May 12, 2022. Maddie's LifeFunder campaign to raise money for essential care can be found at https://www.lifefunder.com/maddie.*]

The Overview

The de Garay family of five—with Maddie being the youngest—lives in Cincinnati, Ohio. In late 2020, one of Maddie's two older brothers heard about Pfizer's clinical trial at Cincinnati Children's Hospital from a teenage friend who had participated, along with one of his parents (see "Clinical Trial Sites and Their Vested Interests"). Influenced by these and other friends, Maddie's parents allowed their three children to enroll in the Pfizer study. Maddie was 12 years old at the time. For their part, the parents signed on for the AstraZeneca trial, but it was halted before they received any vaccinations.

Within hours of receiving her second Pfizer dose, Maddie began experiencing a cascade of alarming symptoms that have left her confined to a wheelchair and eating through a feeding tube. Nine months in, Maddie's mother Stephanie described her daughter as "trapped in a body that doesn't work remotely close to the way it did before."[315] Stephanie, who once had a high-powered career as an electrical engineer—followed in more recent years by fulfilling work as a substitute teacher and special education aide—now spends her time trying to find helpful treatments for her daughter.

Mother and daughter have traveled to Washington, DC, twice (in June and November 2021)[316,317] to speak at events organized by Senator Ron Johnson in addition to sharing Maddie's story at other public events and in numerous interviews in the independent media.[318] Despite the attention from a prominent U.S. senator and others, Maddie's situation has been ignored or downplayed by the FDA, CDC, Pfizer, the mainstream media—and, to a significant extent, also by the hospital that administered her injections. Despite Maddie's nightmare vaccine reaction and its aftermath, board members of Cincinnati Children's had the *chutzpah* to state in their annual report for 2020, "We are especially grateful to the children, young adults, and families who place their trust in our compassionate scientists and clinicians."[319]

Clinical Trial Sites and Their Vested Interests

Cincinnati is home to the Gamble Vaccine Research Center at Cincinnati Children's Hospital, which has played a frontline role as a COVID vaccine clinical trial site for Pfizer as well as AstraZeneca. To this day, the Center's website has a pop-up that states, "Volunteers Needed for COVID-19 Clinical Trial: We are looking for children and adults to participate in a COVID-19 vaccine clinical trial."

As an employer, Cincinnati Children's is a juggernaut, with more than 16,000 employees. Almost a third of the hospital's employees work in research. In 2020, the hospital received over $240 million in external funding, of which 75% were federal—and mostly National Institutes of Health (NIH)—dollars. The hospital also enjoyed a 35% increase in philanthropic funding for research that year compared to the previous year, largely because "so many researchers . . . shifted their focus to defeating the novel coronavirus."

One of Cincinnati Children's focal areas of research seems to be to inventory a wide range of intrusive data about babies. In 2020, the federal government awarded nearly $7 million to "collect entire sets of genomic data" from African American mothers and infants and combine them with other data in electronic medical records; another federal grant awarded in 2021, worth $8 million, will follow mothers and children for a decade, collecting anthropometric, brain imaging, and biospecimen data as well as information about fetal development, family and medical history, and social, emotional, and cognitive development.

Sources:

Cincinnati Children's. Research annual report: a message from Mark Jahnke, Nancy Krieger Eddy, PhD, and Michael Fisher. https://www.cincinnatichildrens.org/research/cincinnati/annual-report/2020/board

Cincinnati Children's. Research annual report: by the numbers. https://www.cincinnatichildrens.org/research/cincinnati/annual-report/2020/by-the-numbers/external-funding
Cincinnati Children's Hospital Medical Center. $75M federal grant includes Cincinnati Children's in drive to increase diversity in genomic research. PR Newswire, Jul. 1, 2020. https://www.prnewswire.com/news-releases/75m-federal-grant-includes-cincinnati-childrens-in-drive-to-increase-diversity-in-genomic-research-301087267.html
Engel L. Cincinnati Children's awarded $8M federal research grant. *Cincinnati Business Courier*, Oct. 7, 2021. https://www.bizjournals.com/cincinnati/news/2021/10/07/cincinnati-childrens-awarded-8m-research-grant.html
Gamble Vaccine Research Center. https://www.cincinnatichildrens.org/service/g/gamble

Warning Signs

Prior to the Pfizer clinical trial, Maddie was a strong, healthy, athletic 7th grader who did well in school, woke up on time, and never had to be prodded to do her homework—as well as a "social butterfly with an infectious sense of humor." Excelling at any sport she tried, she even played for a time on her older brother's indoor soccer team. A deeply caring young person, Maddie regularly babysat and would help neighbors with young children "just to help."

Although generally healthy, all three younger de Garays and both parents take medication for ADHD. Before her injury, Maddie took her medication (Vyvanse—a central nervous system stimulant)[320] from Monday through Friday "for school" but not on the weekends or during the summer. All five have experienced tics and minor eye problems that are known and common side effects of ADHD drugs.[321]

Maddie and her brothers received all recommended childhood and adolescent vaccines, including flu shots and the human papillomavirus (HPV) vaccine. Although Stephanie did not notice adverse reactions at the time, she admits she is reevaluating her children's health history—ADHD included—in light of what she has come to learn about vaccine risks and adverse effects. She now believes many children have vaccine reactions, but "their parents don't have a clue" and "the doctors just brush it off."

Evidence strongly suggests that heavy metals in vaccines[322] (including the thimerosal still found in flu shots and in "trace amounts" in other U.S. vaccines),[323] in tandem with other environmental exposures such as fluoride and pesticides, bear significant responsibility for the rise of ADHD.[324] Studies comparing vaccinated and unvaccinated children and adolescents have also highlighted vaccination as a likely ADHD culprit.[325,326]

Looking back to when Maddie was born, Stephanie recalls that the hospital insisted on giving Stephanie a tetanus shot, saying she "needed to do it for her baby." She remembers that the shot "put her over the edge"; Stephanie also had noticeable reactions to subsequent flu shots.

Stephanie remembers that one of her sons suddenly developed allergies and had bloody stools sometime right before his first birthday—the very time period when the vaccine schedule calls for children to receive as many as nine shots against 13 diseases. No one told the de Garays that virtually all childhood vaccines list allergic reactions as a known adverse event,[327] nor that the package inserts for rotavirus vaccine (administered at two, four, and six months)[328] and the quadrivalent ProQuad MMRV vaccine (typically administered between 12 and 15 months)[329] list bloody stools ("hematochezia") among other reported adverse reactions. Two of the de Garay children also had eczema, another condition linked to vaccination in early childhood.[330]

Maddie received the HPV vaccine in May 2019, along with a Tdap shot and a dose of meningococcal vaccine. The doctor who administered Maddie's shot—a primary care physician who, unusually, does make a practice of reviewing vaccine risks with her patients—mentioned the possibility of an adverse reaction to the HPV shot but described it as "rare." In Stephanie's experience, this was the only doctor ever to give voice to such an eventuality. However, Stephanie says her thought process at the time was, "Am I not going to protect my kids because of a 'rare' chance that something will go wrong?" Despite her own experience with the tetanus and flu shots, she "trusted the doctors."

After the three shots in May, followed in October by a flu shot, Maddie, in early November, had a reaction that doctors should have recognized as a warning sign, first developing strep throat and then mysterious hives. However, when Maddie saw an allergist-immunologist, "everything came back negative," so the allergy specialist—not mentioning that hives ("urticaria") are a documented reaction[331,332] to the HPV vaccine and all flu shots—dismissed her with a diagnosis of dermatographia (also called "skin writing"), an uncommon skin condition "in which seemingly minor scratches turn into temporary but significant reactions."[333] Modern medicine acknowledges "infections" and "medications" as suspected causes of dermatographia.

When the COVID vaccines came along, Stephanie says she "did not go looking for a clinical trial" and that it "had not even crossed her mind" to do so until her son heard about Pfizer's trial via peer word of mouth. Influenced by his friend, he asked to participate, and the de Garays' two other children followed suit; the fact that multiple friends (both adults and kids) had taken part in the trials without mishap seemed like "a lot of reassurance that this was safe." Stephanie herself had, around this time, had COVID, and she worried about the possibility of her kids ending up hospitalized, with COVID visiting restrictions preventing her from being at their bedside.[334] All of the de Garays also were attracted by the promise of "being done with COVID" while "helping others," and they "trusted what the government said."

Two additional factors increased Stephanie's trust level. First, lured by Cincinnati Children's Hospital reputation, Stephanie thought, "if, in the slim chance, the rare chance anything happened, that [her children would] be in the best hands."[335] She reasoned, "If you're going to have anything happen, the best time would be in a clinical trial because they would do everything they could to get you better and to figure out why, because that's the whole point of a trial."

Second, Cincinnati Children's consent forms led Stephanie to believe that "the worst thing that could happen was anaphylaxis—and we'll be in a hospital setting where they have Epi-Pens easily at hand." She admits that with a different experimental product, she might have been more cautious about the safety of something so new and untested, but the monolithic media and hospital messaging about the COVID injections persuaded her that for the vast majority of recipients "cold symptoms" would be the worst of it. Stephanie "didn't know about the different types of adverse reactions" that were possible, whether for COVID shots or childhood vaccines—"they don't tell you that." Maddie herself says, "We didn't think anything was going to happen."

The Tipping Point

After her first Pfizer dose (on December 30, 2020), Maddie had short-term fever and swelling and "didn't feel great," but the family assumed this was a "typical reaction." Things went quite differently after her second dose on January 20, 2021. Maddie received the shot late in the day, accompanied by her father, and told him afterwards that it "hurt worse than the first one." In the middle of the night—highly unusually—Maddie asked her parents if she could sleep in their room. The next day, despite her parents' encouragement to stay home, she chose to go to school but got progressively worse over the course of the day—in pain, dizzy, hardly able to walk, but not wanting to summon her parents. After she got home ("barely" making it off the school bus), Maddie and her dad called Stephanie at work and Maddie told her mother, crying and screaming, "Mom! Mom! My heart, my heart! My heart feels like it is being ripped through my neck!"[336]

A post about Maddie on the *Real Not Rare* website,[337] explains what happened in the ensuing days and weeks:

> *"Less than 12 hours after her second dose, she experienced severe abdominal pain, painful electric shocks on her spine and neck, swollen, ice-cold hands and feet, chest pain, tachycardia, pins and needles in her feet that eventually led to the loss of feeling from her waist down. She also experienced blood in her urine from 7 tests over 3 months, mysterious rashes, peeling feet, reflux, gastroparesis, vomiting, and eventually the inability to swallow liquids or food, dizziness, passing out and convulsions, tics, the inability to sweat, swollen lymph nodes in her armpits, urinary retention, and heavy periods with clots of blood, decreased vision, tinnitus, memory loss, mixing up words, extreme fatigue, and sadly more."*

Cincinnati Children's had instructed the de Garays to log Maddie's reactions for seven days after each dose using a tracking app. However, the app did not allow for any open-ended responses, only supplying a short list of predesignated symptoms, with none of Maddie's most serious symptoms on the list. As a result, "there was no way to report all of Maddie's symptoms."[338] From the beginning, Stephanie kept her own meticulous log of the problems her daughter was experiencing.

When the worried de Garays called Cincinnati Children's, as instructed by the clinical trial protocol, the clinical trial personnel told them to take Maddie to the ER, which performed an ultrasound and blood tests. As Maddie later recounted, "I think some [of the blood tests] came back abnormal but they didn't say anything about it, obviously, because they don't want their vaccine to be having problems."[339] When the de Garays later tried to obtain Maddie's blood test results, the hospital refused to release all of them.

During that first ER visit, the family was sent home after five hours, only to return a couple of days later as Maddie's pain and other symptoms continued to worsen. Again in Maddie's words, "[T]hey kept checking for stuff, and stuff would come back, but they would dismiss it and they would say, 'It's not a big deal.'" As Stephanie puts it, "in this trial, it was all about making the results look good."

The Diagnosis

In 2021, neuropsychiatrists began ramping up their use of a dubious cover term, "functional neurological disorder" (FND),[340] as a gaslighting trick to explain the tsunami of severe neurological reactions being reported all over the world following COVID vaccination.[341] Psychiatrists conveniently define FND—a condition they describe as being "at the interface of neurology and psychiatry"—as a diagnosis for individuals "with 'medically unexplained' sensorimotor neurologic symptoms."[342] FND is claimed to be the second most common reason for neurological outpatient visits.[343] In the context of the COVID injections, neurologists hasten to add that the "close development of functional motor symptoms after the vaccine does not implicate the vaccine as the cause of those symptoms."[344]

Despite Maddie's panoply of 35-plus symptoms, Cincinnati Children's grabbed hold of an FND diagnosis early on. Stephanie states that this premature verdict closed off the possibility of the hospital ordering additional tests that might have pointed the way to beneficial early interventions. Noting, for example, that Maddie's condition resembles the multisystem inflammatory syndrome in children (MIS-C) that has been described as a complication of COVID,[345] Stephanie says timely MIS-C treatments (steroids and IVIG) can make a world of difference, but because of the FND diagnosis, these were not options made available to Maddie.[346] MIS-C has been reported in COVID-vaccinated children as well, including an eight-year-old boy who died seven days after his second dose of Pfizer's shot.[347]

The psychiatric nurse who formalized the FND diagnosis noted in Maddie's chart that he had "collaborated" with the clinical trial's principal investigator (PI) in reaching the diagnosis.[348] Stephanie points out that this represents a substantial conflict of interest: "You don't 'collaborate' with the clinical trial's PI when you're treating a child who got injured in the trial!"

Clinical trial PIs do not file adverse event reports with VAERS. Instead, they are supposed to report adverse events to the drug manufacturer (in this instance, Pfizer), which is then charged with reporting onward to FDA (see "Murky Ethical Oversight"). Although she and her husband tried repeatedly to share details about Maddie's deteriorating health with the clinical trial team, to this day, they do not know what Pfizer actually reported to FDA. They have asked for that information and gotten no answer.

FND was entered into Maddie's chart just one day before Pfizer submitted its EUA request to FDA for Maddie's age group. Stephanie believes that the EUA for 12- to 15-year-olds could never have gone through if Maddie's full adverse reaction had been properly communicated to FDA. After the EUA for adolescents went through in May,

Stephanie filed a VAERS report herself, wanting Maddie's injury to be publicly documented.

In July 2021, the *New England Journal of Medicine* published a study about Pfizer's clinical trial results in 12- to 15-year-olds—with Cincinnati Children's PI as lead author.[349] The publication's authors coolly declared Pfizer's shot to have a "favorable safety profile," claiming, "There were no vaccine-related serious adverse events and few overall severe adverse events." FDA's extension of Pfizer's EUA to children as young as 12 was made on the basis of those data.[350]

Before the "functional neurological disorder" terminology became fashionable, FND's historical precursor had been "conversion disorder"—seen as "an entirely psychological disorder in which repressed psychological stress or trauma gets 'converted' into a physical symptom."[351] Although this "all psychological" explanation has been considered obsolete since the mid-2000s, Stephanie says that psychologists at Cincinnati Children's "grilled" Maddie and her parents and "kept trying to find something traumatic" in her "very normal life" to justify the FND diagnosis. They tried zeroing in on family events such as Maddie's father's recent recovery from cancer or the recent loss of two of Maddie's uncles, but Stephanie points out that "those are not reasons for an FND diagnosis," nor do they explain "why Maddie supposedly developed FND twelve hours after receiving the vaccine."

At the end of the grueling psychological evaluation, Stephanie concluded, "It's almost like they tried to drive Maddie crazy." Stephanie has talked to other vaccine-injured individuals who share similar tales. "That's what they do—they take everything you say and put it in the charts to build a story." When she later obtained and reviewed Maddie's charts, she discovered that there were also many things the hospital personnel *didn't* put in her daughter's chart—especially when the information did not support the contrived FND diagnosis.

Medical Experimentation

Compounding the problems triggered by the COVID vaccine, Maddie also had adverse reactions to a number of medications and medical

Murky Ethical Oversight

Proper reporting of adverse events, serious adverse events, suspected adverse reactions, and unexpected adverse events is critical to the analysis of risks and benefits in clinical trials of new drugs and biologics. Yet, though the FDA provides some guidance on clinical trial adverse event reporting, the Association of Clinical Research Professionals (ACRP)—an organization with members in 70 countries dedicated to "promoting excellence in clinical research"—asserts that "an inadequate level of reporting by [clinical trial] stakeholders, especially of the more serious [adverse event] variety," widely prevails.

The ACRP, whose members "agree to adhere to a professional code of ethics defining essential ethical behaviors for clinical research professionals," has outlined numerous problems with adverse event reporting in clinical trials:

- Most discussions of adverse event data collection challenges focus on postmarketing surveillance and overlook flaws in clinical trial adverse event reporting.
- Determining whether adverse events have occurred involves "asking the right questions." In Maddie's Pfizer trial, the only adverse events the family could report were a handful of less severe symptoms on a preselected checklist, none of which captured what happened to Maddie.
- There is almost no published literature on the *quality* of reporting of clinical trial adverse events and serious adverse events.
- It is unclear whether PIs even understand their reporting responsibilities for adverse events. FDA data indicate that failure to report adverse events—which constitutes "inadequate subject protection"—is a "continual clinical investigator deficiency."

- Underreporting of various types of adverse events—by both PIs and sponsors—leads to "biased evidence and possibly serious consequences for patients."

In addition, HHS itself admits that not only is there is no common definition of the term "adverse event" across government and non-government entities, but HHS's own regulations ("45 CFR 46") for the protection of human research subjects do not even define or use the term.

In a European analysis of adverse event reporting in randomized controlled trials, British and German authors agreed that "the collection, reporting and analysis of [adverse event] data in clinical trials is inconsistent." Among the problems these authors noted were the widespread failure to report whether withdrawals from trials were due to adverse events, and failure (in 84% of studies) to report repeated adverse events.

Sources:

45 CFR 46. https://www.hhs.gov/ohrp/regulations-and-policy/regulations/45-cfr-46/index.html

About ACRP. https://acrpnet.org/about-2/

Neuer A. A fresh take on the adverse event landscape. *Clinical Researcher*. 2019 Feb 12;33(2). https://acrpnet.org/2019/02/12/a-fresh-take-on-the-adverse-event-landscape/

Office for Human Research Protections (OHRP). Reviewing and Reporting Unanticipated Problems Involving Risks to Subjects or Others and Adverse Events: OHRP Guidance (2007). https://www.hhs.gov/ohrp/regulations-and-policy/guidance/reviewing-unanticipated-problems/index.html

Phillips R, Hazell L, Sauzet O, Cornelius V. Analysis and reporting of adverse events in randomised controlled trials: a review. *BMJ Open*. 2019;9(2):e024537.

U.S. Department of Health and Human Services; Food and Drug Administration. *Guidance for Industry and Investigators: Safety Reporting Requirements for INDs and BA/BE Studies*. Center for Drug Evaluation and Research and Center for Biologics Evaluation and Research; December 2012. https://www.fda.gov/media/79394/download

interventions administered afterwards. As Stephanie puts it, "You go in, and they put you on more meds, and it sets off a chain reaction." For example, during one of her hospitalizations, Maddie was given gabapentin,[352] an anti-seizure drug that "can cause life-threatening breathing problems" and, paradoxically, seizures. Maddie reacted badly to the drug within 30 minutes of taking it. "Out of the blue, you have a kid who usually likes to have every light on in the house saying the lights were hurting her eyes; she was seeing spots and couldn't breathe." Maddie had to be on oxygen throughout the night. Stephanie didn't connect the dots between the reaction and the drug until she later looked back at Maddie's chart.

On a different occasion, Maddie was given Lyrica,[353] another anti-epileptic and anticonvulsant also used for nerve pain in adults. Lyrica's serious side effects include "possibly life-threatening" allergic reactions—including swelling, trouble breathing, and hives—and "suicidal thoughts or actions." In Maddie's case, Stephanie says, "that's when her seizures got really bad."

Maddie had MRIs of the spine and the brain, using gadolinium-based contrast agents, and had bad reactions to both (see "MRIs and Gadolinium"). Gadolinium is a heavy metal, and Stephanie now believes Maddie is allergic to metals. After the brain MRI, Maddie could no longer hold up her head, and after the spinal MRI, her walking became much worse.

Day-to-Day Health Impacts

In the case of Maddie and other vaccine-injured children and teens, one of the biggest challenges, according to Stephanie, is that "injured kids are so much different than injured adults," making it hard to find treatment resources. Stephanie quickly learned that Cincinnati Children's and other leading pediatric hospitals were not very helpful, because the majority of such hospitals are running COVID vaccine trials. "If that is happening at the hospital, you might as well not even waste your time."

The logical next step, Stephanie explained in a January 2022 interview, would be to find a pediatrician or pediatric specialist willing and able to help, but there, too, are barriers:

> *"[T]hen you have the added layer of the current stigma with anyone being able to talk about the COVID vaccines, so then that narrows down your pool of people that are willing to take care of her even*

MRIs and Gadolinium

Gadolinium is a heavy metal toxic to humans. When used as an MRI contrast agent, it theoretically has been rendered "safer for use in the body" through a chemical process that allows "healthy kidneys [to] expel the chelated gadolinium out of the body through urine before it can cause toxic reactions." In 2017, however, FDA issued a "drug safety communication" warning that gadolinium-based contrast agents (GBCAs) "are retained in the body." Specifically, FDA—which has approved eight different GBCAs—announced that gadolinium can remain "in patients' bodies, including the brain, for months to years after receiving these drugs."

FDA's warning was late out of the gate, with a 2020 publication in the *Journal of the American College of Radiology* stating, "It has long been known that GBCA administration leaves behind residual gadolinium," including "uptake in various tissues, including bone, kidney, and the brain," even in individuals with an intact blood-brain barrier and healthy kidney function and liver clearance. Researchers coined the term "gadolinium deposition disease" (GDD) in 2017, hypothesizing that GDD could manifest after a single infusion of GBCA "or after multiple administrations in a dose-dependent manner."

While insisting that "the benefit of all approved GBCAs continues to outweigh any potential risks," FDA noted in its 2017 warning that it had collected reports of "adverse events involving multiple organ systems." Children are among the subgroups that FDA considers to be "at higher risk for gadolinium retention."

Sources:

FDA drug safety communication: FDA warns that gadolinium-based contrast agents (GBCAs) are retained in the body; requires new class warnings. FDA, Dec. 19, 2017. https://www.fda.gov/drugs/drug-safety-and-availability/fda-drug-safety-communication-fda-warns-gadolinium-based-contrast-agents-gbcas-are-retained-body

"Gadolinium." https://www.drugwatch.com/gadolinium/

Harvey HB, Gowda V, Cheng G. Gadolinium deposition disease: a new risk management threat. *J Am Coll Radiol.* 2020;17(4):546-550.

> *further. So she has a very, very small window of being able to get medical care, and she has to do it under guise of darkness and quietness and delicateness because the word 'COVID vaccine' is involved."*[354]

Vaccine-injured adults for whom mainstream medicine falls short often seek out health providers working outside the conventional insurance system—assuming the patient has the not-insignificant resources to pay out of pocket. Relatively few of those providers work with children, however. Stephanie comments, "I can't even explain how hard it is to see your child suffer while watching doctor after doctor refuse to help her."

Financial Impacts

Before Maddie's injury, the de Garays were in good shape financially, with a mortgage and one car payment but no other debt. Stephanie describes herself and her husband as "very frugal." She admits she "can't warrant spending tons of money on material stuff" and is not embarrassed to shop at thrift stores, while her husband is "handy" and able to fix things himself.

Initially, Cincinnati Children's paid some of Maddie's medical expenses as a clinical trial participant—but only after the de Garays hired a lawyer to get the hospital to step up and pay. Later, the hospital had them fill out extensive paperwork for Medicaid long-term care (not "regular Medicaid"), which now covers some costs. The de Garays also have health insurance, but the co-pays are considerable and insurance does not cover many of the tests indicated for Maddie's situation.

Stephanie notes that a vaccine injury such as Maddie's comes with a number of less obvious costs that people may not take into account when they assess vaccine risks. These include costs associated with travel to seek out treatments and, even more significantly, loss of income. During 2021, Stephanie was a contract worker, so when she took time off for her daughter's appointments or care, she didn't get paid. Currently, she is on leave and not working at all. They also had to install a wheelchair ramp at home. A co-worker with a daughter in her 20s who had a similar reaction to Maddie's (cause unknown) organized a nonprofit called "Because of Brandi" and donated a wheelchair to Maddie while she waited four months to get hers.

After the de Garays figured out that finding a hospital to help Maddie was a dead end, they tried connecting with providers via telehealth but found that it did not work well in their situation. As Stephanie says, "Maddie really needs the in-person care." Eventually, they moved on to doctors who don't take insurance—"the only ones who can help"—who not only must be paid out of pocket but tend to prescribe many supplements not covered by insurance.

In March 2022, Maddie and her mother temporarily moved to the West Coast, remaining through the month of June. There, they found two doctors—"saints," says a grateful Stephanie—who were willing to help Maddie pro bono. The doctors visited Maddie's rental apartment daily because Maddie needed to be monitored every day. If the two providers decided to try Maddie on a new supplement, they generously purchased it themselves first, only having Stephanie pay for it if it turned out to be something beneficial for Maddie. All of this was financially helpful, but the de Garays still had to cover the rental car and their temporary apartment, which in their case included the cost of making the apartment wheelchair-accessible (AirBnB rentals generally are not accessible). They furnished the apartment with an air mattress, a cheap futon, some items discarded at their building dumpster, and others from Goodwill.

Maddie has a LifeFunder campaign that by June 2022 had raised about $32,000.[355] Stephanie comments that while this may seem like a lot of money, "there's a lot to pay for." As one example, when Maddie's body "swelled out" due to medications, they had to keep getting her new clothes.

When Stephanie and Maddie attend events to share Maddie's story, the events' organizers generally pay the cost of their flights and hotel, "but you still have to pay for food, baggage, Uber, and so on—everything adds up." (As a side note, Stephanie notes that it is "not easy to do Uber with a wheelchair.")

Stephanie comments that although they could sell their house, orchestrating the selling of a house takes a lot of time and energy. Right now, all of their time and energy go to Maddie's treatment. Additionally, they have made their current house wheelchair-accessible for Maddie; were they to move, they would have to find a house with that feature.

The de Garays did not file a claim with the CICP because their lawyers, for various reasons, advised against it. Moreover, CICP compensation—not only highly unlikely but with a cap of $50,000—"wouldn't even come close to covering Maddie's medical expenses" nor Stephanie's lost income.

Social Impacts

Maddie's injury has affected the whole family. Among other impacts, the stress has affected the performance of Maddie's brothers—formerly straight-A students—at school, and both have stopped asking their parents for anything. Both brothers now "work all the time"—one works two jobs—and they talk about looking forward to summer so that they can work more. This is very hard for Stephanie to watch. As she puts it, "the problems you thought you had before pale in comparison."

Early on, Stephanie was able to connect with others injured by COVID shots, including Brianne Dressen,[356] a mother of two who incurred injuries comparable to Maddie's in the AstraZeneca clinical trial. Dressen founded REACT19, a science-based nonprofit "offering financial, physical, and emotional support for those suffering from longterm Covid-19 vaccine adverse events globally."[357] Its REACT-CARE component includes a donor-funded grant program to help get injured individuals "on the path to healing."[358] Dressen has become a close friend of the de Garays. Over time, their small band of vaccine-injured individuals grew "bigger and bigger." Stephanie often gets messages on Facebook now from the parents of other injured children who want to know what to expect and where they can get help.

The family received a lot of support from the school where Stephanie was working—meals, gift cards, and many other kindnesses. Family and friends, too, have been supportive, but Stephanie says she generally "does not have the energy" to talk to friends who she used to touch base with regularly. The de Garays were also regular church-goers, but when Ohio's pandemic policies shut down their church, they lost that direct connection with their church friends.

Stephanie says it was "by accident" that she became a public figure. Early on, she posted requests for prayers for Maddie on a closed Facebook page; when a relative asked to make the posts public so as to share the information with his own church network, the posts "went viral"—and unfortunately also attracted a large number of "mean comments" from people accusing Stephanie of propagating "conspiracy theories." Stephanie now says, "I'm not trying to impress. I'm just telling a story, and it doesn't need to be well polished. I'm just being authentic." Maddie's involvement in a prominent clinical trial has given them more of a voice.

Maddie Today

Stephanie describes Maddie's current muscle control as being "like that of a baby." Maddie cannot lift her legs or hold up her neck. Maddie says that she "loves to sleep" and does a lot of it. Cognitively, Maddie also sometimes mixes up words even while believing that she is saying the correct word.

Maddie is reliant on nasogastric (NG) feeding and has experienced challenges with feeding formulas. In November 2021, she was having issues with one brand of formula (Kate Farms), so doctors switched her to another brand (Abbott's Elecare Junior). The hospital sent them home with two cans of the Abbott product, and the rest were delivered via home care. All of the containers subsequently turned out to be part of the widely publicized Abbott recall,[359] initiated due to bacterial contamination and the production facility's "failure to maintain sanitary conditions." The Abbott formula was a disaster for Maddie, whose stomach "blew up like she was pregnant"; by late December, she was violently vomiting and getting significantly worse. After yet another ER visit (Maddie has had about a dozen), they switched back to the Kate Farms formula, leaving Maddie with an even more "messed-up" stomach.

During Maddie's stay on the West Coast, the goal was to get her stabilized to the point where she would be able to do treatments at home. The doctors were able to significantly improve her overall health with IVs through a peripherally inserted central catheter (PICC) line for fluids, nutrition, natural supplements, and vitamins. When Maddie arrived in March, her liver and kidneys were not functioning properly at all. The doctors were able to stabilize her to the point where the family is now comfortable continuing to work with them virtually. However, Maddie is still unable to feel from the waist down, cannot walk, has cervical instability, and remains unable to swallow. She also has gastroparesis; vision problems; abdominal, back, and neck pain; and other neurologic issues.

Maddie's 14th birthday was in May. Stephanie reports that "her doctors generously paid for her to go home for a week, knowing she was homesick and needed to see her father, brothers, and friends. This was an awesome trip and much needed."

In mid-June, mother and daughter also traveled to New York to see a neurologist recommended by the parent of another child injured by Pfizer's COVID vaccine. The doctor carried out a series of tests (including an electromyogram, a tilt table test, and a punch biopsy); results available thus far are consistent with "severe distal acquired demyelinating polyradiculoneuropathy"—a form of chronic inflammatory demyelinating polyradiculoneuropathy (CIDP). Maddie's test results also point to "small fiber sensory neuropathy and orthostatic intolerance." As of the end of June, the de Garays were waiting for the final test results and hoping for an insurance go-ahead for IVIG, which is FDA-approved as a CIDP treatment.

As founder of REACT19, Brianne Dressen has heard many injury stories. Even so, she says:

> *"But of all the cases, there's a couple of cases that are the top two of massive neglect, massive mishandling by the drug companies, the test clinics, the government, and that's Maddie de Garay's case.... Her doctors gaslit her. They abandoned her. They did everything they could to hide her and to put a label on her, and that child right there deserves way better than what she got, hands down."*[360]

With strong faith, Stephanie says, "I believe God takes bad things and tries to make good out of them. That's why I feel very confident that Maddie is going to recover. Of course, there are days where I'm scared. If you'd told me this story beforehand, I would have said, 'no kid can handle that.' But God chose her because she's strong and gave her extra strength."

As for Maddie, when the de Garays ask one another "What are you thankful for?," her response is, "I'm not dead."

Ernesto "Junior's" Story

[*As told by Ernesto Ramirez, father of Ernesto "Junior," on April 18, 2022. A website in Junior's memory can be found at https://jrsguardianvoice.com/.*]

The Overview

Ernesto Ramirez, an Army veteran and Texan, raised his only biological child, Ernesto "Junior," as a single parent from the time of Junior's birth. Ernesto says, "When Junior was born, the world changed to the best for me"; he took on his parenting responsibilities "with pride,"[361] embracing everything from diapering to day care to school events to coaching Junior's Little League team for seven years—doing everything he could "to make sure Junior had a good life." Ernesto says, "I taught my son how to love God, country, and family." Ernesto himself had never had a father figure; "I did not want my son to go through life without knowing what a father was."

As Junior grew into adolescence, he was "polite and respectful, loved by all who met him," a "wonderful boy, happy and always full of smiles." A star athlete and student, he was Ernesto's pride and joy—the "best son a father could ask for." Father and son spent a lot of time together, sharing a love of outdoor activities like fishing and camping as well as baseball. Ernesto recounts, "When the baseball coach would yell, 'Ernesto's up,' everybody would move back—he was quite a pitcher. It was a heck of a show." Junior had aspirations to be a professional baseball pitcher, but also joined high school ROTC because he wanted to go to college and join the Air Force.

When COVID hit, Ernesto says he "got scared" because at that point, it was "just him and Junior." "They were showing on TV all the time that people were ending up in the hospital, putting them on ventilators and passing away; it seemed like there was no out." As soon as the EUA injections became available, Ernesto took two doses of the Moderna shot "so he wouldn't bring anything home to his son," and he paid attention "when they started announcing that it was safe for teenagers."

Like Stephanie de Garay, Ernesto was alarmed by media reports of hospitals not allowing family members to be with or visit relatives hospitalized with COVID. He worried about Junior ending up in the hospital; "I knew I would not be okay with not being allowed to be by my son's side." That concern, plus the fact of having experienced no adverse reaction to the Moderna shots, and the media's incessant advertising about "how safe it was for the kids" all factored into his decision, in April 2021, to get Junior the Pfizer injection, at that time authorized for ages 16 and up. Ernesto told his 16-year-old son, "Let's get you the shot so we don't have to worry about COVID."

On April 19, 2021, Junior received a single Pfizer dose at a large Texas hospital. At the hospital, a long line of people was waiting to get the shots; Ernesto says, "They were pushing us through like cattle," with no opportunity for meaningful informed consent and only a one-page information sheet.

Five days later, their next-door neighbor—the mother of Junior's best friend—offered to take the two boys to the park to play basketball and then out to dinner. Because of COVID, Ernesto "hadn't let Junior out in all that time." Junior's seemingly uneventful receipt of the Pfizer shot put Ernesto's mind "a little more at ease," and he consented to the outing. A short while later, however, he received a hysterical phone call from the friend's mom, who told him, "There's something wrong with Junior." Junior had collapsed in front of his best friend while playing basketball. Ernesto rushed to the park, where an ambulance was already readying his son for transport to the local children's hospital (not the same hospital where Junior received the shot). Junior died a short time later of what was eventually confirmed to be myocarditis (inflammation of the heart muscle).[362]

As Ernesto says, "I decided to do the right thing and keep my son safe—this turned out to be the worst decision of my life. Suddenly my son was dead, and I was planning his funeral."

Warning Signs

In Junior's case, there were no warning signs. Before he died, he had no preexisting health problems. His father says, "I never had to take him to see a doctor for anything other than regular annual physicals for baseball."

In childhood, Junior received "whatever shots were required for school," with no hint of any adverse reactions. Ernesto had never even known of anyone who had an adverse reaction to a vaccine—and "you never hear it on the news." The fact that Junior had no immediate reaction to the Pfizer shot made it more difficult for Ernesto later to "grab hold" of the fact of his son's sudden death.

The Tipping Point

Since 2021, professional athletes in the prime of life—on average, 23 years old—have been collapsing on the field, after suffering "mysterious" heart problems, including heart attacks.[363] Many of the doctors treating the players are listing "their injuries and deaths as being directly caused by the [COVID] vaccine."[364] As legendary British soccer star and sports commentator Matthew Le Tissier has evocatively described, the collapses—occurring globally wherever teams have required athletes to get COVID injections—are unprecedented in his decades of involvement with sports on and off the field.[365]

Health websites state that individuals who develop myocarditis often have "no symptoms at all," yet they also warn—not explaining how a symptom-free person would even know to "completely avoid all sports for up to six months or longer"—that "the risk of exercise-induced sudden death is real even with mild cases of myocarditis."[366]

Ernesto believes that Junior's minimal "overexertion" at the park was probably the tipping point, resulting in his sudden death.

The Diagnosis

The pediatric hospital conducted an autopsy on Junior but did not tell Ernesto it had done so. When he learned that an autopsy had been conducted, Ernesto asked for a copy of the results, only to be given the run-around for almost three months. Ernesto was able to get a copy only after he threatened to take the matter up with his attorney.

The autopsy recorded a scraped knee, arm, and head from Junior's collapse at the park, but otherwise showed that Junior was in perfect health, with one standout exception—Junior's heart was more than twice the normal size for his age. Later, cardiologist Peter McCullough reviewed Junior's autopsy findings and confirmed the cause of death as myocarditis.

Ernesto then decided to ask for all of Junior's hospital records. The hospital again put him off, asking why he wanted the paperwork and stating that they could "only give it to the parents." When he demonstrated that he was Junior's sole custodian, they told him to come back the next day. He then received what were supposed to be 65 pages of records, but with seven pages missing and other pages blurred and difficult to read.

In August 2021, Ivory Hecker was the first reporter to interview Ernesto.[367] Fox 26 Houston had fired Hecker in June after she publicly accused the network of "muzzling" her coverage of various topics and stated that she was "not the only reporter being subjected to this."[368] On social media, she wrote, "It's important for the public to see that many good truth seeking reporters are left with a choice of sticking to the narrative or serious consequences."[369]

As well as interviewing Ernesto, Hecker sought out others involved in Junior's final days, including the doctor-researcher at the hospital where Junior got his injection, who is involved in clinical trials of COVID shots for children.[370] The doctor attempted to deny that Junior had been vaccinated at his hospital and, according to Ernesto, "backstepped all the way through the interview," even after Hecker showed him Junior's vaccination card and paperwork proving the hospital's involvement. To explain away Junior's death, the doctor then speculated that perhaps Junior had been obese; when Hecker showed him a photo of the fit, healthy teen athlete, he was forced to acknowledge that Junior "looked like a normal 16-year-old kid." By March 2022, this same physician was assuring the media that his county intended to persuade the remaining "200,000 people . . . that have not been immunized" to get COVID shots,[371] and the next month, he suggested that people might "need" the shots as often as every six months.[372] As Ernesto tells it, "He still makes commercials, smiling and telling children to come get vaccinated."[373]

Medical Experimentation

The association between COVID vaccination and heart problems did not come out of left field; well before COVID, researchers understood that vaccination could trigger myocarditis and pericarditis (inflammation of the heart lining) or myopericarditis (both combined). Clinical data and VAERS data have flagged the two conditions as adverse events[374] reported after vaccination against anthrax, Hib, hepatitis A and B, HPV, influenza, Japanese encephalitis,[375] meningococcal illness, smallpox, typhoid, chickenpox and shingles. In 2018, a case report described myocarditis in a six-week-old infant following a DPT shot.[376] Researchers have also pointed out that deriving rates of vaccine-related heart problems from passive data sources like VAERS may well underestimate the true magnitude of the problem; a study of smallpox vaccine recipients, for example, found the incidence of "subclinical myocarditis" to be 60 times higher than the incidence of "overt clinical myocarditis."[377]

There can be no doubt that the incidence of cardiac adverse events has skyrocketed in young people like Junior following the rollout of the COVID shots (see "The Many Faces of Myocarditis"),[378] with evidence suggesting that the new technologies deployed in the EUA injections may be to blame for the dramatic surge in heart problems.[379] A review published in late 2021 identified at least 42 studies reporting cardiac side effects after COVID vaccination, primarily in adolescent or young adult males.[380]

By January 2022, over 3,600 cases of heart disease had been reported to VAERS for persons under 30 who had received EUA COVID vaccines in a one-year period. Before COVID, VAERS had received 23 reports of heart disease per year for the same age group following receipt of any FDA-approved vaccine; this amounts to a 15,600% increase linked to the COVID shots compared to the 31-year period before the shots appeared on the scene.[381] Researchers admit that "the temporal association of the receipt of the vaccine and absence of other plausible causes suggest the vaccine as the likely precipitant."[382]

Participating in Senator Ron Johnson's November panel on COVID vaccine mandates and injuries, Ernesto asserted that Pfizer and FDA knew about the risk of myocarditis in teenagers well before FDA said anything to the public about it, suggesting the agency withheld information vital to the informed consent process.[383] Others, including Dr. Naomi Wolf and the volunteers she has assembled to review Pfizer documents being released by FDA,[384] agree that as regards myocarditis risks in teenagers, there are "grave questions about what the FDA knew and when they knew it."[385] Wolf, who emphasizes that "Adverse events tallied up in the internal Pfizer documents are completely different from those reported on the CDC website or announced by corrupted physicians and medical organizations and hospitals," bluntly proposes that what is underway is genocide.[386]

The Many Faces of Myocarditis

In mid-May 2021, FDA issued an EUA for use of Pfizer's COVID shot in 12- to 15-year-olds,[387] extending the prior EUA granted for ages 16 and up. By May 2022, VAERS had received hundreds of myocarditis reports for the 12- to 17-year-old age group—over 650—with 98% of them linked to the Pfizer injections.[388] As of that date, VAERS also listed a couple dozen reports of myocarditis in 5- to 11-year-olds. CHD's *The Defender* has pointed out that the CDC "uses a narrowed case definition of 'myocarditis,' which excludes cases of cardiac arrest, ischemic strokes and deaths due to heart problems that occur before one has the chance to go to the emergency department." On May 27, *The Defender* also reported noticing "over previous weeks that reports of myocarditis and pericarditis have been removed by the CDC from the VAERS system in this age group."

Listed below is a sampling of the stories published by *The Defender* about heart problems in young people given EUA COVID injections beginning around May 2021:

- **May 26, 2021:** 18 Connecticut teens hospitalized for heart problems after COVID vaccines, White House says young people should still get the shots.
- **June 4, 2021:** 7 U.S. teens developed heart inflammation after second Pfizer vaccine, new study shows.
- **June 15, 2021:** 19-year-old college freshman dies from heart problem one month after second dose of Moderna vaccine.
- **June 15, 2021:** Exclusive: Dad says life "not the same" for 21-year-old student who developed myocarditis after second Moderna shot.
- **June 21, 2021:** Exclusive: Athlete who recovered from COVID facing "very different future" after second dose of Pfizer vaccine triggers myocarditis.
- **June 22, 2021:** Exclusive: Teen who had heart attack after Pfizer vaccine: "I'd rather have COVID."
- **June 23, 2021:** Exclusive: Teen suffers severe heart damage after second Pfizer dose, mother says hospital "clueless" about reporting to VAERS.
- **July 7, 2021:** Teen who had heart attack after COVID vaccine tells RFK, Jr.: "I thought it was safe."
- **August 2, 2021:** CDC study on 12- to 17-year-olds who got Pfizer vaccine: 397 reports of heart inflammation, 14 deaths.

- **August 8, 2021:** 25-year-old develops myocarditis after Moderna vaccine, mother says doctors "downplayed" connection.
- **August 11, 2021:** Exclusive interview: Mom whose 14-year-old son developed myocarditis after Pfizer vaccine no longer trusts CDC, public health officials.
- **October 19, 2021:** FDA delays decision on Moderna vaccine for adolescents citing heart problems, yet Pfizer authorized for teens, despite higher number of myocarditis reports.
- **October 22, 2021:** 17-year-old develops multisystem inflammatory syndrome and myocarditis after Pfizer vaccine, report shows.
- **December 20, 2021:** CDC monitoring 8 cases of heart inflammation in 5- to 11-year-olds who got Pfizer vaccine.
- **December 20, 2021:** 26-year-old's death from heart inflammation "probably" caused by Pfizer COVID vaccine, authorities say.
- **January 6, 2022:** CDC not investigating myocarditis death of 13-year-old days after Pfizer shot, emails reveal.
- **January 11, 2022:** Exclusive: Autopsy confirms 26-year-old's death from myocarditis directly caused by Pfizer COVID vaccine.
- **January 14, 2022:** Myocarditis tops list of COVID vaccine injuries among 12- to 17-year-olds, VAERS data show.
- **February 1, 2022:** COVID vaccines cause "irreparable" damage to kids' brains, heart, other organs.
- **February 11, 2022:** 6-year-old with vaccine-induced myocarditis "unable to walk," as reports of deaths, injuries after COVID vaccines climb steadily.
- **February 18, 2022:** Autopsies show: vaccinating teens for COVID is literally "heartbreaking."

- **March 11, 2022:** 7-year-old died of cardiac arrest 13 days after Pfizer shot, VAERS data show.
- **March 25, 2022:** 664 reports of myocarditis in 5- to 17-year-olds after COVID shots, VAERS data show.
- **March 29, 2022:** Heart damage found in teens months after second Pfizer shot, study shows.
- **May 27, 2022:** Young boy died of myocarditis after Pfizer vaccine, says CDC before signing off on 3rd shot for kids 5-11.

Financial Impacts

In an interview for the CHD.TV show "The People's Testaments" in March 2022,[389] Ernesto recounted a telephone interaction he had with the Federal Emergency Management Agency (FEMA) at the end of 2021. A friend had told him about FEMA's COVID-19 Funeral Assistance program,[390] which offers up to $9,000 for U.S. deaths "attributed to COVID-19" since January 20, 2020. Ernesto's friend had speculated that perhaps FEMA would help cover the costs of Junior's funeral. However, after Ernesto submitted all of the requested paperwork—including being up front that the COVID shot had killed Junior—FEMA denied his application, saying, "It has to say 'COVID' on the death certificate, so we can't help you."

About eight months after Junior's death, just before Christmas, FEMA called Ernesto out of the blue, with the FEMA representative telling him she was calling for more paperwork in support of his application for funeral assistance. Ernesto reminded her that FEMA, already in possession of the full set of paperwork, had denied his application because it would not reimburse for deaths caused by the COVID shots. Ernesto described the 45-minute conversation that ensued:

FEMA representative (after putting Ernesto on hold and then coming back on the line): "Mr. Ramirez, we want you to change your son's death certificate so that it says COVID, and then we can help you financially."

Ernesto: "No, I won't be doing that. I'm not going to disrespect my son in that way."
FEMA representative (after again putting him on hold) repeats request that Ernesto change the cause of death on the death certificate to "COVID."
Ernesto: "No, I will not disrespect my son in that way. I'm not going to falsify documents for financial gain."
FEMA representative (after three-quarters of an hour of back and forth and repeated requests that Ernesto change the cause of death to "COVID," whispering and then hanging up): "I'm so sorry, Mr. Ramirez."

Ernesto speculated that he may have received the call because some people perceive him as "talking too much." Over the course of the conversation, FEMA tried to offer him between $10,000 and $35,000, perhaps figuring, he told CHD, that "this poor father is going to fall for the money." Ernesto noted that "it's a federal offense to falsify documents, and if someone were to get in trouble, it would be me, not the people at FEMA." Ernesto's resolute position: "My son's life was worth way more than that; they'd have to give me the world, and that still wouldn't be enough. They figured I'd want the money. My son's not here. What do I need money for now? Money comes and goes, but Junior is gone." He now wonders, though, "If they did this to me, how many other families have accepted the money?"

After Junior's death, Ernesto decided it was important to "to share his son's story so parents are aware of something he was not," including the fact that taking the shots is "like playing Russian roulette." However, as Ernesto told CHD.TV, he has experienced difficulties in trying to raise money through crowdfunding sites to support travel to speaking engagements. For example, GoFundMe canceled his fundraiser (claiming "prohibited conduct") and forfeited the donations Ernesto had received.[391] As one news account interpreted GoFundMe's action, "For people seeking to mandate vaccination, Ernesto Ramirez Jr.'s story isn't just problematic, it's potentially dangerous."[392] Another fundraiser, at LifeFunder, raised some money but is "not accepting donation [sic] at this point of time."[393] A third fundraiser at GiveSendGo has not been taken down;[394] however, Ernesto says that some people have told him, "I'm trying to donate money but it won't accept my card or let me donate."

Although Junior is dead, Ernesto is very sensitive to the financial plight of the vaccine-injured and their families, such as his friends the de Garays. Like Stephanie de Garay, he encourages the public to donate to REACT19. He notes that the government has billions to spend on overseas military ventures but is offering COVID vaccine victims no financial or medical help.

Social Impacts

Ernesto has paid a heavy price for going public with Junior's story, for example, losing a long-time circle of biker friends who uneasily told him, "you're getting too much attention from the government." During his first trip to Washington, DC, a congressman asked Ernesto if he owned a bulletproof vest and advised him to get one.

On a positive note, the families of other vaccine-injured children have become "like family." Ernesto notes that he has met "a lot of wonderful people," saying, "They're always calling me to check on me, make sure I'm okay." He admits, however, to having "met them for the wrong reasons," stating that under ordinary circumstances, "we probably would have never crossed paths."

Mothers frequently recognize Ernesto when he is out in public and ask if they can introduce their children to him. In one instance, a mother and her teenage twins credited him with saving the teens' lives because they had been planning to get the shot until they heard Junior's story. In another instance, desperate parents contacted him because doctors had persuaded their vaccine-injured daughter that her symptoms were purely psychiatric. Ernesto met with them and helped her see that it was not "all in her head." Ernesto knows of other families in his area with teenage athletes who collapsed after getting COVID shots, but the parents "don't want to speak up."

Junior's best friend—traumatized by Junior's collapse and death—was kicked out of high school for refusing to get a COVID shot. Without proof of vaccination, he also was unable to find a job, so he started his own carwash business. Ernesto asks, "What's he going to do in life now? He still wants an education. They're forcing parents and children to do things that they really don't want to do."

Ernesto Senior Today

One year after his son's death, Ernesto thinks about his son and the loss to society "all the time." He says, "I'm not going to see my son graduate. I'm not going to see him go off to the Air Force. I'm not going to see him married, and I'm not going to see him with my grandchildren." Instead, he visits his son's grave every Sunday, using hand shears to keep the site looking well tended. In November 2021, when Junior would have turned 17, Ernesto bought a cake, sang "Happy Birthday" to his son, and had a "balloon release," but when a friend brought Ernesto gifts the next month for Christmas, he could not bring himself to open even one.

In the immediate aftermath of Junior's death, Ernesto's grief was so deep that he briefly questioned his faith and thought about ending his life. Soon, though, he realized, "I can't do that, because doing harm to myself would mean I couldn't go to heaven, and I need to see my boy again." That let Ernesto know that he "still had some faith inside." He made his peace with God, deciding "It wasn't God that took my son; man killed my son."

Ernesto says, "All I ever thought about was me and my son." He says, "If I knew then just a little bit of what I know now, I would never have risked even taking the chance because my son was everything to me." Even when children seem not to have had any reaction to the COVID shots, Ernesto wants parents to understand that no one knows what the long-term effects of the injections will be.

Telling Junior's story has become Ernesto's mission in life. Wherever he goes, he wears a shirt with a picture of Junior on it that says, "Pfizer vaccine killed my baby boy." He says, "The Lord chose me for this reason, because he knows I'm a fighter and won't give up." Ernesto also pointedly remarks, "I want the world to know that I'm not planning to do myself harm."

Over the second half of 2021, as Ernesto increasingly became a public figure, he spoke in Washington, DC, and at rallies and events all over the country. He says, "I never thought I would be in this position, and I didn't know it would get this big." He receives emails, text messages, and calls from people all over the world: "Australia, Africa, Korea, the Philippines, Poland, everywhere." When he tried contacting the Texas Department of Health and Texas Governor Abbott, however, "nobody wanted to talk to me."

In former times, Ernesto says, he was more combative, but in his newfound public speaking role, he has intentionally cultivated a quieter manner; "I know people will listen to me more if I talk calmly." At an online FDA/CDC meeting where Ernesto spoke along with other vaccine-injured individuals, he was supposed to be the first speaker; however, the group insisted that he go last instead, stating that if he were to go first, they were likely to break down crying and not be able to continue with their own story.

Ernesto says his loss has gotten him "thinking more" than he used to. "For all the wrong reasons, I feel like I've turned into an attorney, a doctor, a scientist. That's why I try to help people." He also notes that as he travels to events, he encounters a lot of children with autism and other vaccine injuries; he says, "It makes me wonder how long this has been going on."

Before he died, Junior was on the brink of being gifted a 1955 Ford truck that had been in Ernesto's uncle's garage for decades. Ernesto and Junior often worked on it together. After Junior passed, the uncle insisted that Ernesto pick up the truck anyway, "because it's Junior's." Ernesto leaves it parked in the driveway all week, only using it on Sundays to go to church and visit Junior's grave. He says, "I can't sell it"; again, "because it's Junior's." When Junior's best friend started his carwash, he asked if he could wash Junior's truck.

After Junior's death, a friend gave Ernesto two dogs. The dogs generally do not bark, but on one occasion, they began barking in the middle of the night and would not stop. Ernesto, a diabetic, realized that his blood sugar had dropped to a dangerously low level—and the dogs had saved his life.

CHAPTER SIX

Adult Vaccine Injuries: The Context and the Stories

CHRONIC DISEASE IN AMERICAN ADULTS: AN INTERNATIONAL EMBARRASSMENT

In early 2020, before COVID, the Commonwealth Fund released a report titled *U.S. Health Care from a Global Perspective, 2019*, subtitled "Higher Spending, Worse Outcomes?" The report compared the U.S. to 10 other high-income countries and also to the average for the 36 member countries of the Organisation for Economic Co-operation and Development (OECD).[395]

Among the report's "highlights," the authors emphasized that although the U.S. spends almost twice as much on health care as a share of the economy compared to the OECD average—as of 2018, this amounted to 16.9% of the gross domestic product (GDP) or more than $10,000 per capita—U.S. consumers of health care get very little bang for the buck. Compared to its 10 and 36 peer nations, the United States has:

- The lowest life expectancy (more than two years lower than the OECD average)

- The highest burden of chronic disease
- Twice the obesity rate
- The highest number of hospitalizations from preventable causes such as diabetes and hypertension
- The highest rate of deaths from chronic conditions—deaths considered "avoidable" or "amenable to health care"—with seven out of 10 deaths attributed to chronic diseases[396]
- The highest suicide rates

An equally somber RAND report from 2017 reported that 60% of American adults live with at least one chronic condition, including roughly half of adults in their mid-40s to mid-60s.[397] At the "highest end of the scale," RAND observed, nearly 30 million Americans (12%) live "day in and day out" with five or more chronic conditions, accounting for more than 40% of U.S. health spending. Another analysis determined that treatment of seven common chronic diseases and related productivity losses cost the U.S. economy over $1 trillion annually.[398]

CDC breakdowns of the prevalence in U.S. adults of six common chronic conditions—arthritis, asthma, cancer, cardiovascular disease, chronic obstructive pulmonary disease (e.g., emphysema, chronic bronchitis), and diabetes—provide an even more alarming picture.[399] Table 2 shows the age group breakdowns as of a decade and a half ago; a similar tally would undoubtedly be even worse now.

TABLE 2. PERCENTAGE OF U.S. ADULTS WITH CHRONIC CONDITIONS (AS OF 2008)

Age Group	1+ chronic conditions	2+ chronic conditions	3+ chronic conditions
Age 55 years and over	78%	47%	19%
Age 55 to 64 years	70%	37%	14%
Age 65 years and over	86%	56%	23%

ADULT VACCINATION AND ADULT HEALTH

Clearly, a myriad of factors contribute to American adults' bad health, including poor nutrition, environmental toxicants, and ubiquitous exposure to electromagnetic radiation. However, vaccines cannot be ruled out as one of the pernicious influences, especially given the aggressive push (even before COVID shots) to increase adult vaccine coverage (see "Vaccination Cradle to Grave" in Chapter One).

Consider the Commonwealth report's praise for the high influenza vaccine coverage in American senior citizens. According to the report, more than two-thirds (68%) of U.S. seniors (age 65 and older) get flu shots versus 44% of OECD seniors, shots that the foundation euphemistically calls a "prevention measure."[400] But whether flu shots "prevent" anything—or improve health at all—has been subject to debate for quite some time. In fact, there are numerous reasons to question the massaged flu statistics continually furnished by CDC.[401] Many studies suggest the jabs are more harmful than beneficial,[402] and in fact, adult flu shots are the injury most often compensated by the NVICP.[403]

In a 2011 doctoral dissertation on "medical politics," Peter Doshi (now a senior editor at *The BMJ*) explained that campaigns promoting flu shots rest on a "shaky scientific basis" of vastly overstated claims about influenza's threat to public health, with vaccination policies more often than not shaped by industry.[404] The same year that Doshi wrote his analysis, the IOM published a report evaluating the epidemiologic and mechanistic evidence linking various vaccines to various adverse events. The chapter presenting the IOM's conclusions about influenza vaccines identified almost 300 studies pertaining to over 20 different types of reported adverse events—especially inflammatory and autoimmune conditions.[405] The IOM committee nevertheless decided that only for two conditions—anaphylaxis and oculorespiratory syndrome—did the evidence support a causal relationship between vaccine and outcome ("convincingly," in the case of anaphylaxis); for the remaining adverse events, the IOM remained neutral, pronouncing the evidence "inadequate to accept or reject a causal relationship."

Table 3 shows the conditions for which the IOM issued its wishy-washy verdict—neither admitting nor denying causation.

TABLE 3. INFLUENZA VACCINE ADVERSE EVENTS FOR WHICH EVIDENCE, ACCORDING TO IOM IN 2011, WAS "INADEQUATE TO ACCEPT OR REJECT A CAUSAL RELATIONSHIP"
Acute disseminated encephalomyelitis (brain and spinal cord inflammation)
Arthropathy (joint disease, including arthritis)
Asthma exacerbation in children and adults, reactive airway disease episodes (live attenuated vaccines)
Brachial neuritis (shoulder pain)
Chronic inflammatory disseminated polyneuropathy (neurological disorder affecting arm/leg function)
Encephalitis (brain inflammation)
Encephalopathy (brain dysfunction)
Fibromyalgia (chronic pain and fatigue)
Guillain-Barré syndrome
Multiple sclerosis (onset or relapse)
Myocardial infarction (heart attacks)—flu shots "transiently increase the risk of cardiovascular events"[406]
Neuromyelitis optica (inflammation of optic nerve and spinal cord)
Optic neuritis (inflammation of optic nerve)
Polyarteritis nodosa and vasculitis (blood vessel inflammation)
Seizures
Small fiber neuropathy (severe pain beginning in hands and feet)
Stroke
Systemic lupus erythematosus
Transverse myelitis (spinal cord inflammation)

Published literature has also linked influenza vaccination during pregnancy to miscarriage[407] and inflammatory responses associated with adverse perinatal health outcomes such as preeclampsia and preterm birth.[408]

The safety and benefits of other adult vaccines have also been widely questioned, including vaccines administered to the military. Following the 1990–1991 Gulf War, a variety of researchers reported health problems strongly suggesting a link between military vaccines and "later ill health,"[409] particularly following administration of multiple shots.[410] A decade after the war, a Veterans Affairs researcher acknowledged vaccines used during the Gulf War as a likely "contributing factor" to the excess morbidity observed in Gulf War veterans.[411]

In 2002, the Government Accountability Office (GAO) issued a scathing report that revealed that 85% of Air National Guard and Air Force Reserve members who had received one or more anthrax shots had experienced an adverse reaction, a rate of adverse reactions "more than double the rate published in the vaccine manufacturer's product insert."[412] The GAO report also noted that the anthrax vaccine program had led to a "mass exodus" of trained and experienced personnel refusing the mandated vaccine.[413]

Shingles vaccines offer another egregious example of adult vaccination gone awry, with Merck's shingles product finally pulled from the market in late 2020 due to a surge in lawsuits following widespread reports of vaccine injury. (As an adult-only vaccine not on the childhood schedule and not recommended for pregnant women, shingles vaccines are one of the few not to benefit from the NCVIA's liability protections.) Flu shots and other adult vaccines, including COVID shots,[414] are also known to trigger reactions labeled as "shingles," which may well represent a detoxification reaction to vaccination.

TRAVEL VACCINES: ANOTHER POTENT TRIGGER FOR ADULT VACCINE INJURIES

When Americans travel abroad, the CDC strongly recommends that travelers of all ages get themselves "up-to-date" with "routine" vaccines (listing shots such as DTP, influenza, MMR, polio, and shingles) and also pushes other shots such as hepatitis A and B, rabies (if the traveler expects to be around dogs or wildlife), typhoid, and yellow fever.[415] This can potentially add up to numerous injections over a short period of time.

Although travel vaccines generally attract little attention, in 2019, yellow fever vaccines briefly made headlines after a leading cancer researcher in the United Kingdom suffered fatal organ failure following yellow fever vaccination and word leaked out that this was not an uncommon occurrence.[416] Published literature and other accounts also describe a range of other adverse events experienced in the aftermath of yellow fever shots, including psychosis, heart damage, and various forms of "invasive and disseminated disease in . . . otherwise healthy individuals."[417]

As the following story illustrates, the pile-on of shots administered for travel can set a vaccinated individual's health and finances back for years or decades.

Lisa's Story

[*As told by Lisa Eden on March 29, 2022.*]

The Overview

Lisa Eden is a Grammy-Award-winning opera singer now in her late 40s.[418] She knew at a young age that she was born to sing, but at least twice, her musical education and career were disrupted by severe vaccine-related illness and injuries. Serious illness during her senior year of high school delayed her college attendance by a full year; after college, travel vaccines at age 24 derailed her musical momentum, preventing her, for a number of years, from doing the internships, auditions, and professional networking so essential for a high-level career in opera. Although she has done a great deal to recover, with music continually giving her "the impetus to fight for her health," the experiences have had impacts on her well-being, professional trajectory, and finances that persist to this day.

It was not until more than 15 years after her injury that listening to a chance radio interview with Dr. Andrew Wakefield helped her understand the "enormity of the devastation caused by vaccines," bringing her to the realization that her own injury was not "one in a million" as she had, for years, supposed. It was not until then that she learned about what had been happening for decades with autism and other vaccine injuries.

Lisa is married to a fellow vocal artist. Since COVID, both have suffered from lockdown-related upheaval in the arts as well as the constraints imposed by vaccine mandates.

Warning Signs

Lisa was raised by parents who were aware enough of alternatives to conventional medicine not to go racing off to the doctor "every five minutes." They did not know, however, about the downsides of vaccines. Lisa recalls that when her mother took her to get booster shots at age five, she resisted "like a wild animal"; her mother and the staff restrained her to administer the shots. Lisa instinctively knew there was something wrong about the experience but says, "when your fears are not validated, you give up and stop fighting it."

Not long after those childhood booster shots, Lisa experienced her first "blackout migraine," and migraines have remained a problem throughout her life. She also developed eczema as a child. Both eczema and migraines are known vaccine adverse events.[419,420]

In the early 1990s, in the summer before her senior year of high school, Lisa was accepted onto a dance team that required that she get a physical. As part of the physical, she believes she was given a tetanus shot and other vaccines. Not long afterward, she developed a hard breast lump "that very quickly went from the size of a pea to the size of a quarter." Frightened by the doctor's evasive language ("we can't confirm that it's not cancer"), she ended up having a lumpectomy early in her senior year. After a nurse told her that breast lumps could be caused by animal products, she became a vegetarian and, later, a vegan—decisions that did not end up serving her health.

By Thanksgiving of senior year, still feeling "off," Lisa learned that she had mononucleosis ("mono"). Although modern medicine generally attributes mono to a virus (Epstein-Barr), mono's symptoms—including, in Lisa's case, extreme fatigue, weight loss, and brain fog—are virtually interchangeable with adverse reactions to vaccines such as the Gardasil HPV injection.[421] Lisa suspects that the shots she was given during her summer physical triggered the mono. The mono was so bad that she ended up missing many weeks of school and was not able to attend school full-time again until the spring. Even worse, she missed her college application deadlines and had to defer her plans to study opera and dance for an entire year. She now says that having to take that year off was "a foreshadowing of what was to come."

Living at home and commuting to school, Lisa started college at the University of Maryland, a recipient of a Creative and Performing Arts full tuition scholarship. At the beginning of the school year, she was asked for proof of an MMR booster. At the time, Lisa says, colleges were more "blasé" about vaccines, and an official actually told her she could fill out a religious exemption form, but—somehow convinced that it was the thing to do—Lisa went ahead and got an MMR in early 1994 anyway.

During her college years, Lisa occasionally traveled overseas on church missions. Unafraid to venture to developing countries, she loved the travel and felt like "the world was her oyster." The travels came with more vaccines, however, such as hepatitis A and typhoid, and she continued to feel "not quite right." Still eating a vegan diet, she would often over-exercise "to try to feel better," all the while maintaining a punishing crack-of-dawn school schedule. She managed to complete bachelor's degrees in both vocal and piano performance, graduating magna (3.98gpa) and summa cum laude (4.0gpa), respectively.

The Tipping Point

A year before graduation, in July and August of 1997, the university's student health service gave Lisa two hepatitis B shots, along with another hepatitis A injection. At the time, each dose of hepatitis B vaccine contained 12.5 micrograms of mercury.[422] Studies published by autoimmunity experts in 2012 and 2014 describe a link between hepatitis B vaccination and the development of a handful of "immune-mediated" diseases, such as fibromyalgia (FM) and chronic fatigue syndrome (CFS), "that share a common clinical picture as well as a history of a previous exposure to an adjuvant agent"—conditions that researchers have grouped under the umbrella term "autoimmune (auto-inflammatory) syndrome induced by adjuvants" (ASIA).[423,424,425]

In May of 1998, shortly before graduation and in advance of a summer church mission in Thailand, her university doctor administered a third hepatitis B shot simultaneously with oral typhoid and oral polio vaccines—Lisa later learned that it was absolutely contraindicated to concurrently give oral typhoid and oral polio vaccines due to the risk of "significant adverse reactions."[426] In addition, the doctor prescribed the antibiotic doxycycline, instructing Lisa to begin taking it that same day. Immediately after the appointment, Lisa developed a debilitating migraine and was advised to take Tylenol (acetaminophen). Subsequently, she found out that acetaminophen depletes glutathione,[427] the body's most powerful antioxidant. (One medical blogger describes glutathione as "the body's secret service agent that is willing under any circumstance to jump in front of a bullet to save you,"[428] saying, "Doing anything to deplete this protector of our health is crazy.")

Lisa says, "Did I question any of this? At the time, not at all." Instead, she maintained her travel plans, continuing on to Thailand to work with a refugee organization. She had intended the trip to be a "gift to herself" and a "journey of self-discovery," but instead—about six to eight weeks after her shots and while in Thailand—her health dramatically deteriorated. Initially, she recalls a "really weird" high fever and fatigue as well as unsteadiness on her feet, but she was reluctant to let on to her humanitarian colleagues that she was unwell. Her job required travel into the jungle, and while there, she developed dysentery-like symptoms—featuring explosive diarrhea "the bright-green color of antifreeze." Unfortunately, the rainy season left her stranded in refugee camps with these symptoms for several weeks. By the time she made it back to her Thai home base, she had stopped eating because it was so painful, shifting her into serious constipation and a cascade of related problems; she also experienced shooting pains throughout her back, shoulders, and neck and, one morning, a protracted episode of body tremors, shaking, and trembling that made moving and speaking difficult.

At this juncture, the head of the refugee organization ordered her to the hospital, where it took a week for her condition to stabilize. Unable to determine the cause of her symptoms, the Cambridge-trained Thai physician treating her encouraged her to return to the U.S. to seek further testing, analysis, and care. Lisa now sums it up this way: "My doctor at the University of Maryland set me up for a pretty terrible vaccine injury, having me do all the things we now know increase the risk of having a severe vaccine injury." In addition to the gut-dysbiosis-inducing doxycycline and glutathione-depleting acetaminophen, Lisa believes that many of her symptoms stemmed from the load of heavy metals—aluminum and thimerosal—in her vaccines.

The Diagnosis

Back in the U.S., practically bedridden, Lisa had no job and no health insurance, so she bought expensive insurance out of pocket. However, her health problems—caused by vaccines that her college student health clinic had promoted and subsidized—were considered "pre-existing," making insurance reimbursement "tricky."

Not knowing where else to go, Lisa returned to see the doctor at the student health clinic. When she ventured to speculate that her health problems might have something to do with "those vaccines," the doctor dismissed the suggestion out of hand. In fact, no one in either Thailand or the U.S. ever said a word about a possible link with vaccination, even when Lisa insisted that she had been perfectly well before the vaccines and unwell ever since.

The doctor prescribed even stronger antibiotics and pain medications, and meanwhile proceeded to test her for tropical diseases such as dengue and malaria, but every test came back negative. The physician then tested Lisa "for all the things she had been vaccinated for"—hepatitis A and B, for example—testing her for HIV, even though she had not been sexually active. Next, she sent Lisa to consult with tropical disease and gastrointestinal specialists, but—despite Lisa's jaundiced skin and other persistent symptoms such as acute gastrointestinal imbalances, skin rashes, thinning hair, weight loss, sensory processing issues, and FM—"no one could find anything wrong." One specialist even said, "You got what you deserve for going to a hell-hole like Thailand—some people aren't meant for travel."

In the latter vein, health providers also increasingly implied that Lisa's problems were psychiatric rather than physical—resorting to tried-and-true vaccine injury gaslighting. In desperation, Lisa—who still couldn't work and was paying for everything out of pocket—found a therapist, who providentially turned out to be extremely kind. During one of their first sessions, Lisa was in so much pain that she could neither sit nor stand comfortably, telling the therapist that on a scale of 1 to 10, her pain level was "11." The therapist told her right away, "This pain is real, not psychosomatic," and encouraged Lisa not to put up with doctors who are "dismissive of young women." Instead, she advised, "If a doctor says you're crazy, find another doctor—and don't go to your appointments alone."

Lisa began reading up on conditions like myalgic encephalomyelitis (aka ME or CFS) and multiple chemical sensitivities (also called environmental illness)—for she had developed food allergies and sensitivities of every kind as well as allergies to animal dander, feather pillows, pollen, scents, and more.[429] She concluded that she was experiencing some kind of "toxic overload reaction."

As she tells it now, her healing journey began when she left allopathic medicine, "which is easy to say but hard to do when you have been raised in that paradigm." She also comments that her journey was necessarily both physical and spiritual; "I had to have faith in a higher power in order to seek guidance and a direction for healing." Lisa explains:

> *"The principal takeaway with my injury is that the doctor caused my injury, and her solutions (that is, more antibiotics and pain medications) caused more and more inflammation and harm, until I realized this and got off the pharma/allopathic medicine freight train that was taking me on a one-way trip to an early grave."*

Medical Experimentation

Lisa found her way to chiropractic care and other holistic therapies. Her chiropractor gave her a job doing office tasks, which allowed Lisa to move out of her parents' home. This generous boss also gave her chiropractic adjustments every day and allowed her to take naps at the office when she needed them. With the chiropractic care, acupuncture, colonics, dietary changes, supplements, and more, Lisa gradually experienced some remission of her symptoms, but she continued to feel that she was not getting to the root cause of her health problems. Although she was increasingly convinced of a vaccine connection, she still did not understand the "how."

A breakthrough came when Lisa heard a radio interview with a biological dentist talking about mercury in amalgam fillings and thimerosal in vaccines.[430] She was "flabbergasted—rocked to her core"—and also appalled that people had treated her "like she was crazy." She wondered, "How could these people tell me it wasn't the vaccines when there is mercury in vaccines?" When she made an appointment with the dentist in question and described all the "crazy symptoms" she had been experiencing since getting her shots, he said she was not "crazy" at all and that he "saw it all the time" (see "Sneaky Mercury"). She learned that she had, in fact, been exhibiting classic symptoms of mercury poisoning.

Lisa's first step after this revelation was to have the dentist carefully remove her amalgam fillings, using biological dentistry's cautious amalgam removal protocols. Next, the dentist tested Lisa for heavy metals. Her level of mercury was literally "off the edge of the paper," and she also had elevated lead. For the next six months, she underwent an intense (and costly) regimen of intravenous chelation therapy, which she credits with "restoring homeostasis." From there, she moved on to gentler and more sustainable detoxification measures.

Sneaky Mercury

Mercury, as one group of experts describes it, "is an unusually insidious toxicant that can cause or contribute to most chronic illnesses," with mutually reinforcing effects on different body systems. As these authors explain, mercury "easily eludes detection," with many mercury-poisoned individuals—as was the case with Lisa—having "no idea that their unexplained health problems are due to past or ongoing mercury exposures." For most people, the principal sources of exposure to mercury are vapor from dental amalgam fillings, methymercury in fish, and ethylmercury in thimerosal-containing vaccines (such as flu shots) and dozens of other pharmaceutical products.

Other key facts about mercury highlighted by these same authors—facts unknown to most members of the public—include the following:

- Mercury has "subtle but reproducible effects on emotions." In all likelihood, "a number of problems blamed on character, personality or stress may in fact be caused or compounded by low-level mercury toxicity."
- Mercury has a particular affinity for the brain but also can accumulate in various organs, glands, and tissues.
- Mercury's toxicity can be "amplified" by other toxic metals, including lead, aluminum, and cadmium. "Mercury and lead, in particular, are highly synergistic." Lisa's heavy metals testing showed off-the-charts levels of both mercury and lead.
- Some types of electromagnetic radiation, including radiation from cell phones, may increase the release of mercury vapor from dental amalgam fillings.

In a preface to Robert F. Kennedy, Jr.'s 2015 book, *Thimerosal: Let the Science Speak*, renowned Harvard researcher Dr. Martha Herbert described mercury as a "biochemical wild card," writing: "That mercury is toxic cannot be disputed. To say otherwise is to pick a fight with the periodic table and the fundamental principles of physical chemistry."

Kennedy has strenuously condemned the continued presence of thimerosal in U.S. flu shots given to pregnant women, as well as the ongoing use of thimerosal-containing childhood vaccines around the world.

Sources:
Kennedy Jr. RF. *Thimerosal: Let the Science Speak*. New York: Skyhorse Publishing; 2015
Kennedy Jr. RF. The ongoing thimerosal travesty needs to end. Children's Health Defense, Jun. 12, 2018. https://childrenshealthdefense.org/news/the-ongoing-thimerosal-travesty-needs-to-end/
Russell S, Homme KG. Mercury: the quintessential antinutrient. Weston A. Price Foundation, Apr. 27, 2018. https://www.westonaprice.org/health-topics/environmental-toxins/mercury-the-quintessential-antinutrient/

Day-to-Day Health Impacts

Following her vaccine injury, Lisa spent about six years intensively focusing on healing, but meanwhile, her musical career was not what it should have been. She says, "When I was in my mid-20s, I should have been out there singing for everybody," but she couldn't manage it. Among other challenges, her brain fog made it difficult to memorize music. The opera career pathway, Lisa explains, "is that you are supposed to either do a master's degree or apprentice or go to Europe, but at the time when I should have done those things, I wasn't physically able to." Although she did gradually resume auditioning and doing gigs, she often needed to go home afterwards and "collapse."

As her health finally improved further, she "leapt with both feet into her musical calling, but there had been so many years lost." By the time she finally started feeling significantly better—now in her early 30s—she had passed the cutoff age for the Metropolitan Opera's National Council Auditions.

In the fall of 2013, Lisa was thrilled to become a resident teaching artist at the Los Angeles Opera in southern California. Disastrously for someone with a history of mercury poisoning, however, a compact fluorescent light (CFL) bulb broke in her hands in the condo she had rented. She soon learned that CFLs contain mercury (see "Fluorescent Light Bulb Hazards"). After she informed her landlord of the bulbs' dangers, she asked him to replace them and he broke a second bulb, casually leaving the broken pieces in her trash can and behind some appliances. Her migraines had, at that point, been in remission, but from one a year or less, she went back to having three to five migraines per month. She moved out and, coincidentally or not, her landlord sold the condo.

In 2016, still in California, Lisa heard about SB 277, the legislation that in 2015 eliminated the state's personal belief (including religious) exemptions to vaccination.[431] It was the first time she had heard about California's legislative assault on bodily integrity and informed choice, but she "instantly and viscerally understood" what was at stake. Shortly thereafter, she attended her first health freedom rally, alternately cheering the speakers and sobbing because it was her "first time among people with similar experiences." In 2017, she began volunteering with Physicians for Informed Consent and has continued to work on behalf of medical freedom ever since.[432]

Financial Impacts

Lisa conservatively estimates that she has spent well over $100,000 out of pocket over the 24 years since her vaccine injury on various therapies that supported her recovery, prevented relapses, and now help her stay healthy—chiropractic, acupuncture, Rolfing, colonics, ozone therapy, and more. During the most acute phase, she had multiple weekly and monthly appointments and took numerous supplements, all the while paying insurance premiums for health care she could not use, "not to mention eating organic food, which adds a lot to your grocery bill."

The health practitioners who she regularly sees today charge anywhere from $130 to $275 per session. Lisa's current chiropractor—who has a relative with autism—gives her discounted services out of gratitude for Lisa's health freedom activism, and Lisa sees her once a week, sometimes more. Lisa continues to spend about $200 per month on supplements. "In the end though," she says, "it is still cheaper than the allopathic model, where you pay and pay and pay only to stay sick."

Lisa notes that after having expensive health insurance for two decades as a freelancer, she switched to "Obamacare" because she qualified for a small subsidy. Over the years, however, her premiums skyrocketed; she also learned that although she had a Platinum Plan (the highest-premium plan, designed to offer "generous cost-sharing"), the doctors in the plan

Fluorescent Light Bulb Hazards

The green energy movement and the U.S. government push CFLs as energy-savers but rarely mention that the bulbs contain mercury. The EPA does acknowledge that when a bulb breaks in the home, "some of this mercury is released as mercury vapor" and continues to be released "until it is cleaned up and removed from the residence." EPA recommends handling CFLs "carefully to avoid breakage." According to the Occupational Safety and Health Administration (OSHA), fluorescent bulbs can also contain "small amounts" of liquid mercury.

For workers involved in handling fluorescent bulbs, OSHA advises the following in the event of "accidental" breakage, or when crushing/recycling machines are broken or opened for servicing:

- Avoiding contact with broken glass
- Not doing cleanup with brooms or vacuum cleaners because they will spread the mercury (unless the vacuum cleaner "is specifically designed to collect mercury")
- Physically isolating areas where workers are processing fluorescents
- Ensuring that the machines are equipped with proper air filtration systems (but even so, "there is still potential for mercury vapor to be released")
- Monitoring mercury air levels
- Providing respirators under some circumstances
- Supplying personal protective equipment "such as coveralls, booties, gloves, face shields and safety goggles to prevent skin and eye contact"
- Providing disposable or reusable protective clothing "to prevent workers from tracking mercury home"
- Conducting medical and biological monitoring, including urine tests and exams "focusing on the eyes, skin, respiratory system, nervous system and kidneys"

- Offering other worker protections consistent with facilities operating hazardous waste treatment, storage, and disposal facilities

Sources:

Occupational Safety and Health Administration. Protecting workers from mercury exposure while crushing and recycling fluorescent bulbs. n.d. https://www.osha.gov/sites/default/files/publications/mercuryexposure_fluorescentbulbs_factsheet.pdf

U.S. Environmental Protection Agency. What to do if a compact fluorescent light (CFL) bulb or fluorescent tube light bulb breaks in your home. Apr. 7, 2014. https://www.epa.gov/sites/default/files/documents/cflcleanup20120329.pdf

were subpar—"you would never select them if you had a choice." One day, her Obamacare insurer canceled her policy without notice, falsely claiming non-payment. For five months, Lisa tried to reinstate the plan, all the while continuing to pay her premiums, but without success. She later learned that the insurer had purged hundreds of Obamacare-subsidized customers without warning or cause.

Social Impacts

When Lisa was first injured, support from family and friends was not readily forthcoming. Professionally, she "lived a double life," never disclosing her health status because she was afraid it would provoke discrimination in the job market.

In 2018, Lisa finally told a close friend she had known personally and professionally for a decade about her vaccine injury and subsequent experiences with the health care system; her friend "laughed in her face," saying, "You can't be serious." When the friend accused Lisa of implying that "all doctors are in a conspiracy," Lisa corrected her: "No, I'm saying they don't know any different."

In 2019, Lisa moved from California to New York, only to discover that New York was repealing its vaccine religious exemption[433]—she was arriving to "exactly the same as what she had just left behind." She quickly plugged into the vaccine-risk-aware community in that state, sharing her experiences.[434]

Lisa has other friends, formerly supportive, who got COVID shots and told Lisa she should get one—despite her medical history. One friend also indicated that she did not want Lisa to share information about the shots' risks. Lisa says that the only people she can have "honest conversations" with are her husband and friends in the health freedom community.

Lisa Today

To this day, Lisa has persistent rashes and migraines. Fortunately, the feverfew plant has helped her get the migraines down to one a month rather than three to five monthly episodes.

Lisa's experiences have led her to the conclusion that "We are literally paying doctors to poison us." She believes there are downsides, never discussed, to many of the drugs and health technologies that conventional medicine touts as "miracles"—which she calls "dead medicine." She also notes that patients are not given true informed consent when entering into immune-system-altering or permanently life-altering procedures, citing mercury amalgam fillings and synthetic hip replacements as examples.

Financially and career-wise, COVID and related mandates have been devastating. As vocal artists in New York—where Broadway and the opera were shut down for months—Lisa and her husband were left without work. Knowing something of the underbelly of past epidemics such as HIV/AIDS—a sordid history outlined in detail in Robert F. Kennedy, Jr.'s best-selling book, *The Real Anthony Fauci*—and familiar with Dr. Fauci's efforts to rebrand myalgic encephalomyelitis (a neurological disorder caused by brain swelling) as "chronic fatigue syndrome," Lisa knew something was awry as soon as she heard of Fauci's involvement in COVID (see "More Fauci Obfuscation").

Lisa's doctor agrees that she should never get another vaccine and she has a medical exemption. When auditions slowly started to come back in New York around August 2020—nowadays carried out by video—Lisa found that she was "literally one of the only singers who was unvaccinated." Her artistic community of fellow singers—proud to "embrace

More Fauci Obfuscation

In her 2014 book *Plague*, scientist Judy Mikovits, PhD tells the story of her search for the truth about the disease originally named myalgic encephalomyelitis (ME). Subsequently "rebranded" as chronic fatigue syndrome (CFS), it is now referred to as both: ME/CFS. As the term "encephalomyelitis" clearly indicates, the condition's hallmark is an encephalitic brain injury.

The NIH says that ME/CFS "lacks a universally accepted case definition, cause, diagnosis, or treatment." As Mikovits and *Plague* co-author Kent Heckenlively describe, however, the fact that the first cases of ME/CFS emerged in 1934 in a couple of hundred doctors and nurses who were probable guinea pigs for an experimental polio vaccine—and the very "polio-like" nature of the illness they experienced—raises important questions about vaccination as a trigger.

Officials like NIAID's Dr. Anthony Fauci have never seen fit to honor or address those questions. In fact, for decades, individuals afflicted by ME/CFS have considered Dr. Fauci "the Darth Vader of the ME/CFS movement" for his role in sending ME/CFS research "into oblivion" and encouraging the pernicious notion, which persists to the present day, that the disease is psychosomatic rather than a brain injury. Lisa's cascade of symptoms started with blinding migraines and neck pain, yet for years she had to battle the medical rejoinder that her problems were psychiatric. In March 2022, a group called "doctors with ME" testily emphasized:

> *"ME/CFS is not Medically Unexplained Symptoms (MUS), Perplexing/Persistent Physical Symptoms (PPS), Functional Neurological Disorder (FND), Pervasive Refusal Syndrome (PRS), Fabricated or Induced Illness (FII), Pathological Demand Avoidance (PDA), Bodily Stress Syndrome (BSS), Bodily Distress Disorder (BDD) or eating disorder.* ***ME/CFS is not***

> ***"functional" or psychosomatic, and these psychological constructs do not apply to ME/CFS*** [emphasis in original]."

The truth is that the sudden, debilitating, and physically rooted nature of the illness has been evident since 1934. As told in *Plague* (p. 69):

> *"Headaches experienced by the [doctors and nurses] were described as being 'of a character and severity never previously experienced by the patients.' The patients also suffered from muscle twitching, nausea, and vomiting. Other complaints included irritability, drowsiness, stiffness in the neck or back, photophobia, constipation, and tremors."*

In *Plague's* foreword, medical journalist Hillary Johnson notes that ME remains "the most common chronic disease most people have never heard of until they acquire it," citing an estimated 20 million cases worldwide and at least a million in the U.S., a number that "exceed[s] the number of patients with breast and lung cancer, AIDS, and multiple sclerosis combined."

Sources:

Heckenlively K, Mikovits J. *Plague: One Scientist's Intrepid Search for the Truth about Human Retroviruses and Chronic Fatigue Syndrome (ME/CFS), Autism, and Other Diseases*. New York, Skyhorse Publishing; 2014.

Johnson C. Does Anthony Fauci finally get it about chronic fatigue syndrome (ME/CFS)? Health Rising, Apr. 28, 2021.

ME/CFS: what psychiatrists need to know. Doctors with ME, Mar. 21, 2022. https://doctorswith.me/me-cfs-what-psychiatrists-need-to-know/

Underhill R, Baillod R. Myalgic encephalomyelitis/chronic fatigue syndrome: organic disease or psychosomatic illness? A re-examination of the Royal Free epidemic of 1955. *Medicina (Kaunas)*. 2020;57(1):12.

'differences'"—shockingly began to ostracize the unvaccinated, treating the lack of a vaccine card as something "worse than leprosy." Medical exemption or no medical exemption, a de facto "blackout" on unvaccinated performers prevails.

For part of 2020, Lisa fled New York for North Carolina, and in October 2021, "totally disenchanted," she moved to Florida to care for her father. The mandates, she says, are "literally keeping me from my calling in life." But she is clear on the importance of standing up for freedom:

> *"We must have the right to choose how we take care of our own bodies. To pretend to have the authority to even legislate such a thing is the ultimate brainwashing. No governing body has the right to infringe upon bodily integrity; whether it is explicitly written as a protection into a constitution or not, body sovereignty is fundamental to human freedom."*

COVID VACCINES: TAKING ADULT VACCINE INJURY TO NEW HEIGHTS

Back in 2005, researchers raised an important but generally neglected question: Do flu shots, touted as "prevention," actually translate into reduced deaths in the population they are most supposed to benefit? After reviewing decades of influenza vaccination data—during a period when vaccination coverage in the elderly went from 16% to 65%—the researchers asking this question reached the officially unpalatable conclusion that there was "no evidence that flu shots reduced death rates" in the elderly, describing a "vast disconnect" between their findings and policy-makers' energetic flu shot recommendations.[435] Fifteen years later, another study not only reiterated that point, understatedly observing that "Current [influenza] vaccination strategies prioritizing elderly persons may be less effective than believed at reducing serious morbidity and mortality," but also revealed, for elderly men, an 8.9% *increase* in all-cause mortality and a 26.5% *increase* in deaths from pneumonia and influenza.[436]

With the experimental COVID shots, these questions are perhaps even more relevant and urgent, particularly when it comes to the shots' impact on all-cause mortality. With damning data emerging from the "numbers don't lie" insurance industry, authorities are having a hard time dodging the murderous implications.

In early 2022, a series of disclosures—not only from the insurance industry but from financial experts parsing insurance data as well as from the CDC and government whistleblowers—revealed shocking correlations between excess mortality and the widespread rollout of COVID shots and boosters:

- For 2021, the nation's fifth largest life insurance company, Lincoln National, "reported a 163% increase in death benefits paid out under its group life insurance policies" for adults aged 18-64.[437] Assuming an average death benefit of $70,000, this would represent 20,647 deaths of working adults "covered by just this one insurance company . . . at least 10,000 more deaths than in a normal year for just this one company."
- In the second half of 2021, a OneAmerica insurance company executive described an "unheard of" 40% increase in the death rate for working-age 18- to 64-year-olds compared to pre-pandemic levels.[438]
- For 18- to 49-year-olds over the one-year period ending in October 2021, the CDC likewise acknowledges a 40% rise in excess deaths.[439]
- The fall of 2021 saw an 89% spike in excess mortality in 25- to 44-year-olds (the Millennial generation), bluntly characterized by market analyst Edward Dowd as "democide" (death by government).[440]
- In the military, senior flight surgeon Lt. Col. Theresa Long has been one of several whistleblowers testifying in court to astronomical surges in miscarriages, cancers, and neurological problems after the military's COVID vaccine mandates went into effect. Long testified to having been contacted by a high-level officer, in a major violation of whistleblower protections, "and told not to discuss her findings regarding the explosive military medical data in court."[441]

Nor are deaths the only outcome displaying an undeniable temporal relationship to the COVID shots' rollout. In June 2022, Dowd described data from the U.S. Bureau of Labor Statistics showing an explosive increase in the disabled population in 2021;[442] from a stable 29 million Americans reported as disabled every year for the previous five years, the ranks of the disabled suddenly expanded by three million—a 10% increase.[443] The actual number could well be much higher because many disabled individuals do not receive disability payments. OneAmerica's insurance executive also reported significantly higher short- and long-term disability claims in the second half of 2021.

The following two stories barely begin to convey the physical and financial misery being endured by adults disabled by COVID shots.

Mona's Story

[*As told by Mona Hasegawa on March 21, 2022.*]

The Overview

Before her vaccine injury in the spring of 2021, Mona Hasegawa was a recently divorced mother of two in her early 40s, a Canadian transplant living in New Jersey with her 11- and 22-year-old daughters and her partner. At the time, Mona had transitioned out of jobs as a nursing home aide and ATM bank technician and was trying to figure out "next steps" in her work life while relishing temporarily being a stay-at-home mom to her daughters. She particularly loved going on bike rides with her 11-year-old and taking her on special outings to the zoo or New York City. After just one dose of the Pfizer vaccine—which Mona got after being told she could not otherwise visit her parents in Canada—she is now disabled, in a wheelchair, and in chronic pain.

Warning Signs

As a child, Mona had "every single vaccine" and never experienced any problems. As an adult, she had taken tetanus shots and one flu shot when it was required for her nursing home job. She had never known anyone with a vaccine injury and thought that the worst that might happen with the COVID shots would be "temporary flu-like symptoms." Other than having slightly high blood pressure, Mona was in good health.

The Tipping Point

Mona received one dose of the Pfizer injection on April 24, 2021. She had wanted to visit her parents and other relatives in Canada, including her father who is on dialysis, and she was told she could not visit him nor stay with her parents unless she got a COVID vaccine. In addition, she had been hearing on the radio—"over and over"—that everyone should get it and that it was "safe and effective."

Mona knew other people who had already gotten COVID shots, seemingly uneventfully. They had all taken the Pfizer shot, so she thought "it must be safe." She drove on purpose 40 minutes away to a CVS drugstore to get the Pfizer vaccine specifically. She says, "I was kind of excited about getting it, and looking forward to not having to worry about COVID anymore."

About two hours after receiving the shot,[444] Mona was having dinner at a local restaurant with her partner and daughters. When she went to the restroom with her oldest daughter, her "mind suddenly froze" and her brain "couldn't tell her legs to move." She had to ask her daughter for help to get up. Perplexed and shocked by her sudden physical incapacitation, Mona asked to leave the restaurant right away. She says, "Surprisingly, even at this point, it wasn't in my head yet that it could be the shot."

Over the next couple of days, Mona developed a variety of debilitating symptoms. Her legs, in particular, felt extremely heavy—"like elephant trunks." She also experienced severe lower back pain. She went to the hospital, but they said nothing was wrong with her and "brushed it off."

Over the next two weeks, her symptoms persisted and multiplied. In addition to feeling extremely weak, she developed seizures and tremors. She tried using a cane, but it wasn't enough. At one point, she had to crawl on her hands and knees to get into her house.[445] Inside, when she needed to go to the bathroom, her oldest daughter had to wrap her in a sheet and drag her because her daughter wasn't strong enough to carry her. At this point, Mona had not yet acquired either a wheelchair or a portable commode.

All this while, Mona was "still trying to be a good mom." A couple of weeks after the shot, she took her daughters to the mall; there, something "sucked the energy out" of her, and she collapsed. Her daughter called 9-1-1, and the ambulance took her on a stretcher to a different hospital from the one she had been to a couple of weeks previously. At this point, Mona was thinking, "this is the vaccine," especially after the emergency medical technicians asked her questions about whether she had recently changed any medications. At the hospital, she insisted it was the vaccine, telling them the full story of what had been happening since the injection. However, after doing x-rays and an MRI, the hospital staff said they "didn't see anything" and that it was "impossible" for it to be related to the COVID shots. Although a neurologist briefly checked in on her, he did not come back. Mona says now, "They didn't take time with me." At that point, she could not walk, but the hospital nevertheless discharged her and said there was nothing they could do for her.

The Diagnosis

Before discharging her, the hospital recommended that Mona see a rheumatologist and neurologist as an outpatient. However, "it takes months to get appointments with specialists." As Mona puts it, this meant "more time with me suffering before I was able to get any appointments." When she finally was able to see a rheumatologist, she was given steroids, for which Mona is grateful; she believes her legs could have become paralyzed if she hadn't taken the steroids.

Following an episode of heart palpitations and seizures, Mona checked herself into yet another hospital. However, when she told them she had been injured by Pfizer's COVID vaccine, they "weren't too happy to hear that." She remained in the hospital for a week as they did "this test and that test"—a spinal tap, another MRI, more x-rays, and lots of blood tests—again reaching no conclusions.

When Mona asked one of the doctors to report her injury to VAERS, he refused, saying "it was too early to know if the vaccine could cause this." Mona spoke with someone directly at VAERS who said they would submit the report for her, but she now can't get into the system. "Locked out of her own report," she is unable to add new symptoms to her own VAERS dossier.

Her lengthy list of symptoms includes chronic venous insufficiency,[446] a condition that occurs when the valves in the leg veins stop working effectively. This causes blood to pool in the legs. She also has spinal issues for which it has been suggested she see an orthopedic surgeon. In addition, she was diagnosed with a ministroke—a transient ischemic attack or TIA—after the left side of her face began drooping.[447] Doctors view TIAs as a potential red flag for future strokes.

Mona has experienced considerable gaslighting from the medical profession, despite doctors' sworn Hippocratic oath to "refrain from causing harm or hurt."[448] At one of New York's largest and most famous teaching hospitals, a doctor called her a "conspiracy theorist" for even daring to suggest that she had experienced a vaccine injury. This made Mona scared to look for help and made it hard for her to trust doctors. At one point, she got "so tired of jumping from doctor to doctor" that she "needed a break."

Medical Experimentation

Mona's symptom picture also includes extreme spikes and sudden drops in blood pressure. Based on conversations with other vaccine-injured individuals and her own research, Mona suspects she may have a condition called postural orthostatic tachycardia syndrome (POTS), a blood circulation disorder characterized by "an uncomfortable, rapid increase in heartbeat" when going from lying down to standing up, often accompanied by decreased blood flow to the brain or difficulty keeping blood pressure "steady and stable."[449] POTS—thought to be an autoimmune disorder[450]—frequently co-occurs with other autoimmune disorders, notably ME/CFS, FM, and systemic lupus erythematosus (SLE).

Many young people injured by the Gardasil HPV vaccine are intimately familiar with POTS,[451] with case reports confirming a strong temporal relationship between the two.[452] In an analysis of nearly 1000 serious adverse event reports submitted to the Danish Medicines Agency (2009–2017) on behalf of adolescent girls and young women following

HPV vaccination, researchers identified the term "POTS" in anywhere from 4% to 30% of reports, while the combination of "headache," "dizziness," and either "syncope" or "fatigue" (hallmarks of POTS) was present in 43% of reports.[453] Eager to exonerate HPV vaccination from any possible blame, however, the study's pharma-funded researchers concluded that "a non-trivial proportion of adolescent girls will experience symptoms after vaccination due to chance alone"—for good measure adding that it is "not uncommon" for teenage girls to have headaches, dizziness, sleep disorders, and stomach aches. At the Mayo Clinic (which counts Gardasil's manufacturer Merck among its "principal benefactors," along with COVID manufacturers AstraZeneca and Pfizer),[454] researchers have gone to similar lengths to downplay temporal associations between HPV vaccination and POTS, despite the hypothesized causal relationship.[455]

Notwithstanding the Mayo researchers' disingenuous statements, vaccination is, in fact, a scientifically accepted cause of "de novo POTS."[456] However, leading autoimmunity experts Yehuda Shoenfeld and Lucija Tomljenovic, who have extensively studied vaccination and vaccine adjuvants as triggers of autoimmune disorders, note that the risk of POTS and related syndromes "may vary between different vaccines."[457] Before the COVID shots, HPV vaccines were the standout culprit, particularly when compared to other vaccines (such as those for meningitis and chickenpox) given to adolescents and young adults. Now, POTS is fast becoming a common diagnosis among COVID-vaccine-injured adults like Mona (see "Life-Changing Vaccine-Induced POTS").[458] One published case report closely echoes Mona's experience, with a 42-year-old male experiencing symptoms of fatigue, headache, and myalgia within 24 hours of taking one dose of Pfizer's vaccine and, within six days of the jab, presenting at a clinic with POTS-like symptoms.[459]

In a survey conducted by REACT19 with over 500 individuals "suffering a wide range of persistent symptoms" after one or more doses of COVID vaccine, two-thirds of respondents (65%) had only received one dose of vaccine, primarily the Pfizer or Moderna mRNA shots.[460] Fatigue (81%), brain fog (68%), dizziness (55%), persistent headaches (41%), tachycardia (36%), and other symptoms consistent with POTS and ME/CFS were among the top reported symptoms, with nearly all respondents experiencing serious problems within one week of vaccination.

Day-to-Day Health Impacts

Over time, Mona found that she could no longer hold her head up without a neck brace and could no longer walk even short distances. When, through social media, she connected with other vaccine-injured individuals, one of them introduced her to a doctor who listened to her story and prescribed ivermectin. The ivermectin made a big difference, helping her hold her head up and walk short distances. Not all of her medical team agreed with her taking the drug, however; one doctor said, in front of his residents, "A doctor should never give ivermectin."

At some point, Mona's insurance company would no longer cover the ivermectin. Mona now has to pay for it out of pocket. Because of the expense, she has tried not taking the drug, but when she stops, she "declines very quickly."

Life-Changing Vaccine-Induced POTS

Robert F. Kennedy, Jr., and attorneys from the national law firm of Baum Hedlund Aristei & Goldman have filed a series of lawsuits against Merck, alleging that the Gardasil manufacturer knowingly concealed adverse events such as POTS and other life-changing conditions associated with its vaccine, often after just a single dose. Extreme disability and pain are common refrains among the young people involved in the HPV vaccine lawsuits, many of whom previously were healthy athletes and academic superstars. One young competitive swimmer and nursing student with POTS, ME/CFS, narcolepsy, small fiber neuropathy, and autonomic dysfunction describes herself as "a shell of what I used to be."

A single dose of Pfizer's COVID shot had a similarly devastating impact on a young woman building a career as an air traffic controller, who did not want the shot but got it to keep her job.

Within 15 minutes of receiving the injection, she experienced arm and chest pain. Over the following months, she developed dizziness, shortness of breath, and memory issues, as well as "twitching, nerve pain, fatigue, high blood pressure, high heart rate, palpitations, lightheadedness, a feeling of vertigo and migraines." Like Mona, most of the doctors she encountered (17 in seven months) "seemed more concerned with assuring [her] it wasn't from the vaccine" than with treating her, telling her it was a "coincidence that the symptoms developed immediately after the vaccine" or that her sudden and very serious health problems were "stress-related." She obtained a POTS diagnosis only after reading about it on her own and visiting a physician who specializes in POTS, who "knew right away that [she] had POTS and understood [her] struggles with previous doctors, because all his POTS patients go through the same things."

Sources:

Children's Health Defense. 25-year-old woman a "shell of what I used to be" after taking Gardasil vaccine. *The Defender*, Jan. 5, 2022.

Children's Health Defense. Young man sues Merck, wants accountability for injuries caused by Gardasil HPV vaccine. *The Defender*, Mar. 11, 2022.

Nevradakis M. Exclusive: 29-year-old's career came "crashing" down after Pfizer COVID vaccine injury. *The Defender*, Jun. 16, 2022.

Financial Impacts

Mona has had great difficulty "finding the right doctors who will take this situation seriously." Moreover, as so many of the vaccine-injured report, the doctors who will listen are not covered by insurance. One doctor whom Mona would like to see costs $5,000 out of pocket—for her, a prohibitive amount.

Mona has health insurance, but there are many things insurance does not cover; "a lot of things bounce back." In Mona's case, insurance will not cover special tests or some MRIs, and it "even rejected the wheelchair and walker." Even with the appointments covered by insurance, the co-pays are financially challenging. It was recommended, for example, that Mona get physical therapy four times a week, but with a $20 co-pay, that would cost her around $320/month. Unable to afford the co-pays, Mona is not getting the physical therapy she needs.

Before the injury, Mona's household's financial situation was "livable," with no debts. Now, she has "thousands and thousands of dollars of debt," with many in collection. Five to six medical bills arrive in the mail every day and Mona finds it difficult to open them. She puts all of the envelopes in a bag that, at this point, is stuffed full. For the sake of her daughters, she now receives government food assistance and other voucher assistance, but this is hard for her: "I hate to ask for money—this is not who I am." When her daughter launched a crowdfund campaign, it only raised $280. She loves to paint and has thought about selling paintings or having people bid on her paintings so that she "can offer something in return."

As the one-year statute of limitations for filing a CICP claim approached, Mona planned to file a claim, while not holding out much hope for compensation. She is aware that CICP does not compensate for pain and suffering, even though the pain and suffering that she, her partner, and her children have experienced have been as bad "as the injuries themselves." She would like to see COVID vaccine injuries get added to the NVICP, which she views as "the only thing that the vaccine-injured have," but she generally acknowledges that "there's no help financially when it comes to vaccine injuries—you're on your own."

Social Impacts

The impact on Mona's household has been significant socially as well as financially. Her partner has been "supportive and active," and also vocal on social media to raise awareness. However, her partner's friends "don't want to hear it," and consequently, she has lost friends. Mona says her partner "is okay with it because she feels like she's doing the right thing."

For Mona, among the hardest things she has had to face are the changes in her ability to be the kind of mother and parent she would like to be. She says sadly, "it is devastating that my daughters have to deal with me in this condition." Mona is no longer able to drive and cannot take her outdoors-loving 11-year-old daughter on excursions or bike rides. When they go to the grocery store, it makes Mona "feel weird" to ask her youngest to push her in the wheelchair and load the shopping cart. It has also been difficult for Mona to get to school events—when her daughter received a school art award, for example, Mona was physically incapable of attending.

Mona's older daughter got a good job around the time of Mona's injury, which involves working a lot of hours. Mona told her daughter that she doesn't want her injury to impact her daughter's life and career, but her daughter is trying to help in whatever ways she can.

Mona's relatives in Canada have not been terribly supportive. One relative cuts her off when she speaks about her injury. Other "old-school Japanese" relatives are reluctant to make waves or "cause problems." Mona says, "Strangers have become my closest friends."

Mona Today

Mona's most recent diagnosis, confirmed via biopsy, is small fiber neuropathy—a condition increasingly viewed as an autoimmune disorder. Small fiber neuropathy involves damage to the small fibers of the peripheral nervous system (PNS). The PNS connects the brain and spinal cord (the central nervous system or CNS) to the organs, limbs, and skin, carrying information about pain, temperature, and other sensory and motor data, and regulating functions like digestion, blood pressure, and heart rate.

With the small fiber neuropathy diagnosis in hand, Mona, like Maddie, would like to undergo IVIG therapy. FDA has approved IVIG for use with several "immune-mediated peripheral nerve disorders," including Guillain-Barré syndrome (GBS) and CIDP. There is growing clinical recognition that IVIG is also a promising therapy for small fiber neuropathy, but it is not yet FDA-approved for that indication. Clinicians have reported dramatic changes in many patients with similar neuropathies, who "improve by 80-90% of their pre-illness level of functioning."[461] They note that "Comparing autoimmune [small fiber neuropathy] to the other peripheral immune-mediated neuropathies that are FDA-approved indications for IVIg has been an effective strategy" for gaining insurance authorization. Mona hopes that this will be the case with her insurer.

Day to day, Mona admits that it has been very hard to cope with her "new life." She says, "I really don't have any quality-of-life right now. My life has turned upside-down. My life is totally the opposite of what it used to be. I am praying that it gets better, and I am hoping they figure out more treatments."

Mona says she does not want others to suffer as she has, so she spends a lot of time speaking with people, including the vaccine-injured. She knows other individuals harmed by COVID shots who are suicidal; "I'm hoping they are strong enough to survive this."

She strongly opposes COVID shots for children. "If I'm having a hard time dealing with this, both physically and mentally, how are the kids going to deal with this?"

Suzanna's Story

[*As told by Suzanna Newell, with occasional input from her husband, on March 28, 2022.*]

The Overview

Suzanna Newell and her husband live in St. Paul, Minnesota. Both highly athletic, the couple enjoyed long-distance biking together, and Suzanna did triathlons—until her vaccine injury. Suzanna has two teenage children, a son and a daughter, by a prior marriage.

A former high school valedictorian, Suzanna's work history included a stint with Teach For America in New Orleans right after college (working in special education) and time in corporate America. The year before her injury, she found her way to a career in corporate social responsibility (helping companies operate "in ways that enhance society and the environment instead of contributing negatively to them").[462]

Soon after the COVID shots became available, Suzanna's husband got two doses of the Moderna vaccine because his job with the school system asked for it. (He then went on to get COVID anyway.) For her part, Suzanna—at age 49—decided in March 2021 also to get the shots, opting for the Pfizer product specifically, "because they were saying it was the most effective." Less than 48 hours after her second dose in April, she began developing serious symptoms that have left her, more than a year later, in chronic pain, with brain fog and unable to work, and often unable to walk.

Warning Signs

As a child, Suzanna got all recommended childhood vaccines, and as an adult, she continued to get annual flu shots—never, to her knowledge, experiencing any adverse reactions. In fact, she was "very pro-flu-shot" because her sister had experienced severe complications from H1N1 influenza in 2014, ending up in a medically induced coma for six weeks and later requiring kidney dialysis. Her sister, who also has neuropathy, has been on disability ever since. Suzanna's mother has significant health issues as well and lives with Suzanna's sister nearby. Before her injury, Suzanna would often go and clean their house or help them in other ways.

At the time of her sister's illness, a doctor sternly said that she "would not have gotten sick if she had gotten a flu shot." Because of that formative experience, Suzanna was not only committed to flu shots for herself but would tell everyone she knew to "get your flu shot." One time, a friend replied that her father had developed GBS from a flu shot. This was the first (and only) time that Suzanna ever heard that something bad could happen post-vaccination, and that "someone got hurt from something I'd been pushing." Nevertheless, she assumed her friend's father was an unfortunate outlier and continued to believe that adverse reactions were "even less than one in a million."

As a child, Suzanna had tuberculosis (TB) at age four. She also had malaria in 1993 when she traveled to Kenya. Aside from those two acute illness episodes, Suzanna was, before taking the Pfizer shots, "healthy as a horse." In contrast to her disabled sister, she was more fit, more active, and "healthier than your average 49-year-old," going to the doctor only for every-other-year check-ups.

Suzanna used her robust athleticism to help people with health problems. She and her husband participated in the "MS 150" bike ride every year (the largest fundraising cycling series in the world), riding 75 miles a day for two days to raise money for multiple sclerosis research. She also took part in annual American Lung Association "Fight for Air" fundraisers that involve climbing 31 flights of stairs. Some of these issues were personal: in addition to Suzanna's childhood TB, her father died of lung cancer and her son has asthma. Suzanna's son also had eczema and reflux as a child, but until her own vaccine injury, she had never considered the possibility that his health issues—which, as we have seen, are common vaccine adverse events—might be related to vaccination.

The Tipping Point

Suzanna followed the news about COVID via a CNN app on her phone. She and her husband now recognize that CNN and other mainstream media put out a powerful "fear narrative" (often, ironically, thanks to Pfizer media sponsorship). Still, Suzanna was not afraid of COVID herself, feeling that her body and immune system could handle it, but she "raced to get the shots" for the sake of her mother and sister. She says, "I was not going to be the one to kill grandma." When she got her shots, Suzanna recalls a feeling of "camaraderie" and celebration—of being "very excited to do my part."

Suzanna got her vaccines at a pop-up clinic set up by her University of Minnesota health network. Within two days of the second dose, she woke up with a swollen lymph node, extreme fatigue, excruciating pain, and a "weird electric shock" feeling; she could tell right away that something was very wrong. She says, "Compared to doing the 'MS 150' bike ride, I had three to five times more fatigue," and her whole body felt like it was "internally vibrating." She also developed a rash that persists to this day. All of her symptoms have been more on the right side, "like a line down the body." She notes that she got both of her shots in the right arm.

Suzanna waited exactly two weeks before going to see a doctor because initially, she thought the problems would go away on their own. Then, as she began feeling "worse and worse and worse," she wondered if she had a sudden cancer or was going through menopause. She knew that something was wrong, but she was "second-guessing" herself.

The Diagnosis

Suzanna went to see her primary care provider, a nurse-practitioner (NP) who had been doing her biannual check-ups. The NP said, "Let's take this one step at a time—we'll do some blood work and an ultrasound, and then go from there." Suzanna believes the NP "likely was aware of what was happening" because, in a subsequent phone conversation, Suzanna asked the NP if she thought the problems could be menopause-related and the NP said—in a very clear and deliberate tone of voice—"No, I do not think this is menopause." Suzanna is grateful for that clarity, because at that point she was still second-guessing herself and asking, "Is it me?"

It took about three weeks for Suzanna to come to the conclusion that "it was definitely the shots." Twin Cities doctors had by then begun conducting a barrage of tests that ruled out cancer but had not produced a diagnosis. Suzanna said to herself, "This happened after the shots. Before, I was totally fine; now, I'm really not fine." At this juncture, however, she still thought her case was an "anomaly." Trying to be a good citizen, she also didn't want to scare anyone else away from taking the shots.

Continuity of care has been a challenge for Suzanna during her vaccine injury ordeal. For example, after just one in-person appointment, her NP abruptly retired. Fortunately, her replacement was a physician's assistant (PA) who Suzanna describes as "amazing"—part of what she refers to as "Team Humanity." He listened to her and did his best to get her help. However, he, too, ended up leaving the practice, and—only a year after her injury—Suzanna had already had to move on to her third primary care practitioner.

Because of the lymph swelling, Suzanna went on to get an MRI of her neck and then was referred to an ear, nose, and throat (ENT) specialist. When the ENT told her he was "not that concerned about the lymph growth because it doesn't look that big," Suzanna was relieved, but at the same time, she still knew "something's really wrong with me." One indicator of things having gone significantly awry was her blood pressure, which previously had tended to be stable and on the low side but had become "super erratic, going really low and really high." Her heart rate displayed the same roller-coaster effect.

By May, things had gotten so bad that Suzanna went to the ER, but again, no one seemed able or inclined to help her. With encouragement from her husband, she then sought a second opinion at the Mayo Clinic, assuming that Mayo "would want to help figure out what was wrong" but discovering that this was not necessarily the case (see "Mayo Revealed"). Mayo was not forthcoming about the results of her tests.

That same month, Suzanna reported her injury to VAERS, but like so many others, she found the system to be "not user-friendly" and difficult to update. In addition, she reported her injury to Pfizer. When Pfizer sent paperwork to her PA essentially asking him if he thought her health problems were related to the shots, he answered "yes."

Mayo Revealed

The Mayo Clinic has long acknowledged Pfizer and other pharma giants as some of its "principal benefactors" (annual giving at the $1 million to $10 million level). Unlike at many other nonprofit hospital systems, COVID has been good for business. In the first quarter of 2021, the press reported that Mayo had "clos[ed] out 2020 on a strong note," signaling a "trend of strong revenue and income performance." And by the end of 2021, Mayo was reporting "big income gains" for the year—a 14% jump in revenue—after "relatively modest increases in previous years." During 2021, Mayo received 11% more grant and contract income and 36% more "benefactor" income compared to 2020. It also administered over half a million COVID vaccinations and performed over a million COVID tests.

In 2021, Mayo implemented a COVID vaccination requirement for all employees other than those able to obtain medical or religious exemptions, firing about 700 workers in January for not meeting its conditions. Nine lawsuits have since been filed by 27 former employees (many of them long-standing) denied exemptions by Mayo; the plaintiffs allege wrongful termination and argue that Mayo failed "to undertake an individual and interactive process for evaluating" religious exemption requests.

Mayo's conflicts of interest are apparent in its publications. For example, five of the 12 authors of an August 2021 preprint comparing the Pfizer and Moderna COVID shots (Puranik et al.) were Mayo Clinic employees, with the study's "competing interest statement" admitting that "The Mayo Clinic may stand to gain financially from the successful outcome of the research." (Nevertheless, a Mayo review board somehow determined that the research was "being conducted in compliance with Mayo Clinic Conflict of Interest policies.")

Mayo's "Get the facts" webpage on COVID-19 vaccines lists only "mild side effects," stating that most "go away in a few days." While also admitting "it's not yet clear if these vaccines will have long-term side effects," Mayo soothingly adds, "vaccines rarely cause long-term side effects." Mayo encourages pregnant women to get the shots, claiming—despite considerable evidence to the contrary—"no serious risks for pregnant women who were vaccinated or their babies."

Sources:

Mayo Clinic Consolidated Financial Report, Years Ended December 31, 2021 and 2020. https://cdn.prod-carehubs.net/n7-mcnn/7bcc9724adf7b803/uploads/2022/02/2021-Mayo-Clinic-Consolidated-Financial-Report.pdf

Muoio D. Mayo Clinic reports $243M operating income, $3.7B total revenue to kick off 2021. *Fierce Healthcare*, May 21, 2021.

Puranik A, Lenehan PJ, Silvert E, et al. Comparison of two highly-effective mRNA vaccines for COVID-19 during periods of Alpha and Delta variant prevalence. MedRxiv, Aug. 8, 2021.

Revenue up, remote care down at Mayo Clinic in 2021. KAAL-TV, Feb. 28, 2022.

Snowbeck C. Former employees suing Mayo Clinic over COVID-19 vaccine mandate terminations. *Star Tribune*, Jun. 23, 2022.

Following her stay at Mayo, Suzanna obtained an outpatient referral to see a Mayo rheumatologist (a specialist who focuses on arthritis, other musculoskeletal conditions, and systemic autoimmune diseases). By this time, she had developed right-sided weakness resembling a mild stroke as well as a strange "pulsating eye." In June, the rheumatologist clinically diagnosed her with idiopathic ("no known cause") Sjogren's syndrome (a condition affecting the eyes and mouth),[463] though blood tests for Sjogren's and lupus had both come back negative. When Suzanna requested a lip biopsy—a test that could confirm or disconfirm Sjogren's—the doctor discouraged her from getting it. Later, when Suzanna obtained her medical charts, she saw that he had charted this interaction as the opposite of what had happened, writing that Suzanna had "refused" a lip biopsy. This and other experiences taught Suzanna that medical charting is often not accurate. (When, in February 2022, Suzanna was finally able to get a lip biopsy at the University of Minnesota, the test confirmed that she does not have Sjogren's.)

Medical Experimentation

About three months after her second Pfizer dose, Suzanna's PA helped get her a referral to a neuro-opthalmologist to assess her "pulsating" pupil. The neuro-opthalmologist turned out to be more interested in her leg pain and sweating than her eye, stating, "I think you have small fiber neuropathy." Based on her own research, Suzanna had already tentatively reached the same conclusion.

Burning and shooting pain (including "electric shock-like pain" like that experienced by Suzanna), sweating, and blood pressure fluctuations are some of the most prominent symptoms of small fiber neuropathy.[464] The syndrome is associated with a variety of autoimmune conditions, and WebMD also acknowledges it as a consequence of vaccination (see "What's in an Excipient? Sometimes, Intractable Pain").[465]

The neuro-ophthalmologist was attuned to small fiber neuropathy because he had it himself and had experienced an exacerbation following a COVID booster. He tested Suzanna using a newer type of eye test but requested that another doctor confirm the diagnosis via a skin biopsy. Researchers agree on the merits of using multiple diagnostic criteria to assess the condition,[466] but when Suzanna requested the skin biopsy in August, the other doctor replied that he "wasn't convinced" and did not agree to order the biopsy until December. Ultimately, the procedure confirmed small fiber neuropathy—a diagnosis that, by that time, Suzanna had had on her radar for six months!

What's in an Excipient? Sometimes, Intractable Pain

The published literature includes a number of case reports of small fiber neuropathy following vaccination. In 2009, researchers concluded "that an acute or subacute, post-vaccination small fiber neuropathy may occur and follow a chronic course," citing the experiences of individuals vaccinated against chickenpox, Lyme disease, and rabies. In 2016, a case study described the "acute onset" of "intractable generalized pain" in a 14-year-old girl nine days after HPV vaccination, agony that persisted unabated "despite various pain medications."

As one possible mechanism for vaccine-associated neuropathy, the authors writing in 2016 posited that it might be triggered by "hypersensitivity to the solvents/adjuvants" in vaccines. In April 2021—the same month as the onset of Suzanna's symptoms—researchers came up with exactly the same explanation for a case of biopsy-proven small fiber neuropathy with onset after COVID-19 vaccination. Pointing out that "Concerns regarding neurologic complications of vaccination are not new," the authors suggested that "an immune-mediated hypersensitivity to the solvent/adjuvant"—in this case, to a compound called polyethylene glycol (PEG)—could be the culprit.

PEG is the coating for the lipid nanoparticle (LNP) "delivery system" used by Pfizer and Moderna to transport synthetic mRNA into recipients' cells. Children's Health Defense began issuing warnings about the dangers of PEG as a component of the Pfizer and Moderna shots even before the shots' emergency use authorization. In February 2021, physicians from Vanderbilt University and Brigham and Women's Hospital concurred, warning in the *New England Journal of Medicine* that PEG could represent a "hidden danger" and noting that the COVID mRNA shots represent the first time vaccines "that [have] PEG as an excipient [have] been in widespread use."

The term "excipient" is generally used to refer to substances other than the active ingredient that are presumed to be "inert." These substances typically constitute 90% of a pharmaceutical product but serve no direct therapeutic purpose; instead, they are included for reasons such as enhanced product stability or more complete drug absorption. The catch, as some authors have pointed out, is that few if any excipients actually meet the criteria of being "pharmacologically inactive, non-toxic, and [not interacting] with the active ingredients or other excipients." Instead, most excipients—including PEG—"have proved to be anything but inert, not only possessing the ability to react with other ingredients in the formulation, but also to cause adverse and hypersensitivity reactions," ranging from "a mild rash to a potentially life-threatening reaction."

Sources:

Castells MC, Phillips EJ. Maintaining safety with SARS-CoV-2 vaccines. *N Engl J Med.* 2021;384:643-649.

Children's Health Defense. These "inactive" ingredients in COVID vaccines could trigger allergic reactions. *The Defender*, Mar. 12, 2021.

Haywood A, Glass BD. Pharmaceutical excipients—where do we begin? *Aust Prescr.* 2011;34:112-114.

Kafaie J, Kim M, Krause E. Small fiber neuropathy following vaccination. *J Clin Neuromuscul Dis.* 2016;18(1):37-40.

Redwood L. 5 questions Fauci and FDA need to answer on Pfizer and Moderna COVID vaccines. *The Defender*, Dec. 23, 2020.

Souayah N, Ajroud-Driss S, Sander HW, et al. Small fiber neuropathy following vaccination for rabies, varicella or Lyme disease. *Vaccine.* 2009;27(52):7322-7325.

Waheed W, Carey ME, Tandan SR, Tandan R. Post COVID-19 vaccine small fiber neuropathy. *Muscle Nerve.* 2021;64(1):E1-E2.

Day-to-Day Health Impacts

Since her injury, Suzanna is "very much a frequent flyer," with some kind of doctor's appointment nearly every day. She describes it as "a game of hot potato between specialists," as she bounces around between neurologists, rheumatologists, cardiologists, gynecologists, allergists, and others. Physical therapy, occupational therapy, and acupuncture are the only things covered by insurance that she has found to be helpful. Whereas in the past, she would work out her stress through exercise, she has had to find new ways to cope, such as meditation.

Suzanna "often has to fight to be believed," and she is still "trying to get doctors to admit that something happened—that it's not her fault and is not genetic." Suzanna also comments, "I have to hold it together so they don't write me off; it doesn't work if I cry in those offices because they tell me I'm 'anxious.' We're being told we need to 'relax.' It's cruel." Over time, she has learned that ideas have to be proposed to doctors "gently." Particularly as a woman and with so much medical gaslighting going on, "you have to be very careful about how to request stuff." She says her corporate America training kicked in and has helped her be "a gentle advocate for herself."

Financial Impacts

Suzanna was previously the household's top breadwinner. Her husband left corporate America a few years ago, and they agreed that he could work at a less stressful but lower-paying job. Not long before getting the Pfizer shots, Suzanna had temporarily transitioned from full-time to part-time status to help her mother and sister. After the injury, she managed to return to work part-time for a while, but when her employer asked her in September 2021 to go back to full-time, she knew that was "not going to work." That month, she went on full-time, short-term disability, which provided 60% of her former salary and health insurance; she is now on long-term disability, with 50% of her previous salary.

The injury has, therefore, amounted to "quite a financial hit." Although the family is managing with the disability and Suzanna's husband's job, she says, "I make lots of choices not to do things that might be good for me because of the cost." Insurance has covered many of Suzanna's health care expenses, but it will not cover some interventions that she suspects would be helpful, such as IVIG. Two different neurologists requested IVIG, but insurance has denied it twice, stating there is "not enough research to prove it will help." In both 2021 and 2022,

Suzanna hit her out-of-pocket maximum right away. She is grateful that acupuncture is covered by insurance because it is "the only thing helping with pain."

Initially, Suzanna was reluctant to dip into their savings, thinking "Who knows how long this will last?" Later, in a desperate attempt to achieve some pain relief, she changed her mind and has now spent tens of thousands. She says, "I am lucky I have that money to spend, but I know there are others like me who don't have that money, and I'm concerned about them." She is immensely grateful for the "kindness of strangers"; for example, she has had providers from out of state spontaneously reach out and gift their services.

In early March, 2022, not long before her one-year statute of limitations was up, Suzanna filed a claim with the CICP. That same month, she received an email from CICP asking for additional information. She recognizes, however, that CICP compensation is a long shot and would, in any case, be inadequate. Like others, she would like to see COVID vaccine injuries transferred over to the NVICP, although she acknowledges that NVICP, too, has major limitations and she admits "that is probably super wishful thinking."

Social Impacts

The first thing that happened to make Suzanna question her assumption that her vaccine injury was an "anomaly" was when she and her husband had an encounter with someone who mentioned that his wife was "really struggling" since getting a COVID shot. Suzanna learned that the man's wife, a fit and active woman in her 60s, had developed carpal tunnel symptoms in both hands "overnight," followed by a muscle disease called polymyositis that "can make even simple movements difficult."[467] In short order, that woman and Suzanna "became each other's support system."

While doing online research, Suzanna also came across the story of a physician who developed symptoms just like Suzanna's within about 15 minutes of her first dose of the Pfizer shot.[468] After reaching out to the doctor, Suzanna joined a private group on social media where she was able to connect with a group of about 200 other people injured by

COVID vaccines. One of them was Brianne Dressen (injured in the AstraZeneca clinical trial and mentioned in Maddie de Garay's story), who experienced first-hand the "total blackout" on COVID vaccine injuries and went on to found REACT19.[469] Members of the online group shared solutions, exchanged information about supplements, suggested possible tests, and provided a sounding board—"a lifeline when you're thinking you're all alone." This was when Suzanna realized she was far from "one in a million."

Suzanna feels that COVID vaccine injuries come with their own unique brand of stigma. For people who want to believe that the injections were—and are—"the way out of the pandemic," it seems to be uncomfortable to hear about her injury, but Suzanna believes it is essential not to "turn away from uncomfortable truths." However, people who are still fearful of COVID illness "don't want to hear that the shots are not the answer." In essence, many people's position is, "Don't tell me anything bad about the vaccine because I believe in it."

Suzanna has friends who have "dropped off," and she doesn't know what they are thinking. "When it's people who know you and disappear because they don't like what you have to say, that hurts." Again, she speculates that some of them may be scared because they took the shot, so they find it preferable to remain in denial or "blame the victim." One friend essentially reduced her injury to statistics, telling Suzanna she "drew the short straw," but Suzanna asked her, "What about the fact that there's no liability and no compensation? How can we require something, and then ditch people?"

The fact that Suzanna's symptom picture is not static but constantly changing (for example, sometimes she can walk, and sometimes she can't) can "make it seem like I'm lying." She says, "When you feel okay some of the time, and then down again, it's easy for people to have doubts." She finds it isolating to have an injury that is hidden by the media and not "socially acceptable like cancer" (although cancer was once highly stigmatized). She knows of COVID-vaccine-injured individuals who have committed suicide due to their suffering and isolation.

Suzanna's injury has resulted in a significant reshuffling of family roles. She can no longer help her mother and sister as she used to do and, even worse, she cannot be the mother that she was accustomed to being. Like Mona, she says, "Having them push me in a wheelchair is the hardest thing—it breaks my heart. I hate that so much." It is also difficult for her to have her kids see her in pain. They, in turn, get frustrated by not being able to do anything to help her when she is suffering. Around the house, her teens are now cooking, cleaning, doing laundry, and doing much more for themselves than they did before. Suzanna also used to more closely monitor their schoolwork but says, "Now they have to self-manage." (She and her husband wryly try to acknowledge a silver lining to this increased self-reliance.) As for her ex-husband, Suzanna describes him as having "gone down the rabbit hole much sooner" than she did, so he has been supportive since her injury.

Suzanna's children each received one dose of the Pfizer shot. The kids now have doctors' notes saying that they should not get any more, but sadly, they experience stigma for being "half-vaxxed," and many of their friends' families remain "ardently pro-vaccine and deeply in the narrative." They cannot get into some places because they are not "fully vaccinated." Early on, they were impatient with these restrictions, but they have come to understand that getting more shots would not be a good idea.

Suzanna Today

Suzanna describes the COVID-vaccine-injured as being in "uncharted territory," noting that it is a "sad and lonely" place to be. She believes the callous official response to COVID vaccine injuries is producing "a pandemic of trauma," with the injured "being traumatized due to our cases being overlooked, misdiagnosed, and hidden." In addition, "When you have a condition where there is no immediate relief and you're not on a straight path to getting better, people get tired of checking in on you." She adds, "I would not wish this on my worst enemy."

Her most debilitating symptom is pain, which she experiences every day to varying degrees, with occasional reprieves and then flare-ups. A stabbing/shooting pain down her right leg makes it difficult to walk, and she has to use a walker or wheelchair most of the time. She also experiences extreme dizziness and sometimes has to "grab onto the walls." Her husband adds that one of the most concerning symptoms he witnesses is chest pain and his wife's spiking blood pressure. Suzanna also finds the brain fog to be "super frustrating," because she "used to be able to come up with things quickly." She says, "I can't do what I want to do anymore." The pain and fatigue make it difficult even to do simple tasks like laundry or chopping vegetables.

Suzanna reports:

> *"I would say the majority of my waking moments are now impacted by my injury since I am in so much pain and because I am now disabled. I have to allow myself 15 minutes more before I go any place because I move so slow and have to get the cane, walker, or wheelchair out. I need a second person to push the wheelchair, and if I take the cane or walker, I have to stop to rest or because of pain. Additionally, I have to drive looking for handicapped accessible entrances, elevators, or parking spaces. It is extremely time-consuming and frustrating. I see the rest of the world walking around able-bodied, and I remember when that was me before the shot. I want that to be me again, but I don't know if hope, want, and hard work at physical therapy is going to be enough to make that happen. I still remain hopeful for improvements, but concerned as time marches on."*

Part of Suzanna's current quest is to find helpful natural healing approaches. However, when she tried breathing techniques, it was "too much for her heart and set off her autonomic nervous system." She found a "brain retraining" program[470] that includes meditation to be moderately helpful—though not an "overnight success"—and she continues to use those techniques as much as she can. Reiki healing and red light therapy are other modalities that have showed promise,[471] with the red light therapy being the first intervention to make a dent in her persistent rash.[472]

She asks that people not turn away but instead "lean in and support and lift," commenting, "We have to support each other." She encourages the public to stop judging whether someone's pain is "justified" or not. Suzanna feels that officials have tried to divide and conquer people and get them to "pick sides," but she prefers to "meet in the middle."

Suzanna is concerned about the vaccines' potential impact on children. "It is one thing for me to be disabled like this in my late 40s, but another thing entirely to damage the lives of perfectly healthy children who we know are at least risk of contracting fatal COVID. They cannot be our lab rats."

At this juncture, Suzanna's advice to others is clear:

> *"Remember that legally, if you get injured by a COVID vaccine, you're not going to have any rights. You will basically be on your own. It is isolating, emotionally taxing, financially taxing, and lonely to be kicked out of regular society. Don't leave the fallen to fend for themselves. There needs to be a safety net—and it should not be the American people. Pharma should be responsible."*

CHAPTER SEVEN

Bringing the True Risks of Vaccination Out of the Shadows

FROM SANITATION AND NUTRITION TO . . . SYNTHETIC MRNA

As recently as the late 1980s—around the time that the NCVIA was engineering a liability-free gold rush for vaccine manufacturers[473]—prominent schools of public health were still teaching students the truth about the dramatic decline in U.S. mortality in the first half of the 20th century, namely, that improvements in nutrition, clean water, sanitation, hygiene, and refrigeration accounted for the successes, not medical measures.[474] Contrary to popular belief, the decline could not have had much, if anything, to do with vaccination because most vaccines did not yet exist,[475] facts plainly revealed by U.S. vital statistics. In short, civil engineers—not vaccine scientists—produced the 20th-century gains in life expectancy.[476] To this day, low-income countries that implement meaningful water, sanitation, and hygiene interventions achieve remarkable health improvements.[477]

These historical facts are essential to keep in mind when being bombarded by incessant and deceptive messaging that humans cannot survive without vaccination—and never more so than now, when government and industry have opened the floodgates to a new "biopharma" (biotech-plus-pharmaceutical) gold rush based on mRNA technology, a technology that thrills vaccine and drug makers because it vastly speeds up the development and manufacturing processes.[478] The lucrative biopharma sector—going even more "mainstream" thanks to the COVID injections—is the fastest-growing segment of the drug industry globally, representing 20% of the worldwide market and displaying an annual growth rate that is more than double that of conventional pharma.[479]

The diligent doctors and scientists who make up D4CE have warned from the outset that mRNA vaccines and the mRNA technology pose a "serious threat to mankind,"[480] notably because of the potential for harmful effects on female and male fertility and pregnancy outcomes.[481,482] As D4CE member Dr. Thomas Binder has explained, mRNA injections can never be safe "because even if a not toxic antigen is chosen, the toxicity of the Lipid Nano Particles, the modified RNA and the auto immune like reaction against the cells, who are coerced to produce and then present this foreign protein on their surface, will be the same." Until the COVID vaccines, these were some of the very reasons why no mRNA injections had ever been licensed.

Nevertheless, having succeeded in foisting COVID mRNA injections on an unsuspecting public, manufacturers and government entities like the NIH are now salivating at the prospect of mRNA vaccines for the conditions being reported as COVID vaccine adverse events: cancer,[483] shingles,[484] heart attacks,[485] and drastic immune suppression (currently being rebranded as "HIV").[486] In addition, manufacturers are gearing up for a new generation of flu shots—mRNA-based—as well as mRNA combination vaccines that they promise will "protect against several different infections at the same time, such as influenza, COVID-19 and other respiratory infections."[487] Ever eager to add more injections to the childhood vaccine schedule, the industry is also eyeing potential cash cows such as a pediatric and adult mRNA vaccine against respiratory syncytial virus (RSV); the last time vaccine makers tried to roll out an RSV vaccine, two babies died and most of the rest were hospitalized.[488]

In 2019, there were 30 candidate RSV vaccines in the pipeline, and in 2021, FDA fast-tracked an mRNA-based RSV vaccine by Moderna.[489]

DUE DILIGENCE REQUIRED

Given the dismal track record of U.S. and global vaccine programs over at least the past half-century, as well as the many alarming features of mRNA and other emerging vaccine technologies,[490] adults contemplating vaccination for themselves or their children must do rigorous due diligence and scrutinize the notion that vaccination is the path toward robust health. Grandparents, too—so often the "bankers" of last resort—must insist on discussing vaccine risks with their adult children so that their grandchildren and their financial legacy can be protected.

Legal dictionaries define informed consent as "the act of agreeing to allow something to happen, or to do something, with a full understanding of all the relevant facts, including risks, and available alternatives," highlighting that "*full knowledge and understanding is the necessary factor in whether an individual can give informed consent.*"[491] Common law, state and federal statutes, the 1947 Nuremberg Code, and the 2005 UNESCO Declaration on Bioethics and Human Rights have all established the necessity of prior, free, and informed consent for medical interventions. As per the Nuremberg Code, persons should be able to exercise "free power of choice, without the intervention of any element of force, fraud, deceit, duress, overreaching or other ulterior form of constraints or coercion."[492] The UNESCO Declaration extended the consent principle from experimentation to "any preventive, diagnostic and therapeutic medical intervention," stating that such interventions are "only to be carried out with the prior, free and informed consent of the person concerned, based on adequate information."[493]

Although these principles represent the cornerstone of modern medical ethics, it has been abundantly clear for decades that the professionals and organizations providing and peddling vaccination do not practice true informed consent and do not disclose "all the relevant facts, including risks, and available alternatives." Although the NCVIA set up a basic stipulation that doctors give out Vaccine Information Statements (VISs) prior to every vaccine dose to inform recipients of risks and benefits, evidence indicates that few doctors even bother, and in any case, the VISs have been so dumbed down as to barely include any information on risks at all.[494] During the mass vaccination events at hospitals and "pop-up" community clinics so popular during COVID—where, as Ernesto put it, personnel were "pushing people through like cattle"—meaningful informed consent has been even more of a phantom notion.

The absence of informed consent also extends to the reality of undisclosed ingredients in vaccines. A 2017 study by Italian researchers uncovered the nearly universal presence in vaccines of "debris" and "micro-, sub-micro- and nanosized inorganic foreign bodies" of "unusual chemical compositions"—ingredients "neither biocompatible nor biodegradable" and not declared in package inserts.[495] Investigators have also identified abnormal human DNA, including genes associated with cancer, in vaccines derived from aborted fetal cell lines—again, without disclosure to vaccine recipients.[496] COVID injections have taken mystery ingredients to new heights,[497] with batches found to contain metallic particles and other contaminants.[498,499] In September 2021, Japan halted the distribution of 1.6 million doses of Moderna vaccine after finding stainless steel and rubber fragments.[500] In April 2022, Moderna recalled almost three-quarters of a million doses in Europe due to contamination by an undisclosed "foreign body."[501]

In securities law, material omissions—that is, "making statements that paint an incomplete or inaccurate picture, and not revealing other material information necessary to present the entire truth"[502]—are a violation of the law and subject to prosecution, fines, and jail time. As discussed above and in Chapter Two, the vaccine informed consent process (to the extent that there is any) is a sham. "Vaccine informed consent" provides no disclosure or statistics related to the financial costs and risks associated with vaccine injury, disability, or death, such as the risk of bankruptcy, foreclosure, or depletion of intergenerational wealth;[503] no discussion of the potentially significant impact on time and resources available to other family members; no mention of the increased chance of divorce; and no attention to the drain on non-financial forms of family wealth such as human, intellectual, social, and spiritual capital.[504] These are material omissions. Dr. Ryan Cole, an Idaho pathologist, urges individuals injured by COVID vaccines to sue their doctor for not giving them genuine informed consent.[505] Cole states, "Any physician giving [COVID shots] and not giving informed consent is putting you involuntarily into the largest trial on humanity ever."

Individuals who decide to take COVID shots—or any other vaccine, for that matter—have a responsibility to make the appropriate provisions for their family, beforehand, should death or disability result. This includes assessing the adequacy of health, disability, and life insurance policies (and obtaining additional insurance if needed); making sure living wills and health care proxies are in place; carrying out estate planning; and making caregiver provisions in the event of disability. For COVID shots, the Solari Report's "Family Financial Disclosure Form for COVID-19 Injections" can help ensure that persons intent on accepting the injections take responsibility for protecting their families from financial disaster.[506]

KEY THEMES

The goal of this book is to shine a light on the risks that government, the media, the medical-pharmaceutical complex, and others willing to play with ordinary people's lives would prefer to sweep under the carpet, describing the enormous price paid by individuals who trusted their most precious asset—their health—to duplicitous and criminal entities.

The nine stories presented herein underscore a number of compelling themes common to the experience of vaccine injury, including the fundamental and cruel reality that the vaccine-injured are ON THEIR OWN in responding to the medical and financial challenges of events that are frequently life-changing—if not fatal. The combination of a medical profession trained to be incurious and indifferent, a legal climate engineered not only to favor industry but to encourage recklessness, and a government infrastructure either complacently tolerant of or willfully causing harm[507]—as the government did in the 1920s and 1930s when it was "intentionally poisoning alcohol to encourage compliance" during Prohibition[508]—has left injured individuals and families sick, bewildered, gaslit, stressed, angry, isolated, and broke.

Some of the other themes highlighted in our stories include the following:

- **Many people appear willing to outsource their critical thinking and defer to what doctors, the legacy media, and puppet spokespersons for "science" tell them.** Parents of children injured as babies and now in their 20s and 30s noted that they were especially reliant on doctors' counsel back when home computers and Internet sources of information had not yet become widespread. In the current moment, respondents also described the powerful influence of the media's fear-mongering about COVID and of the "safe and effective" mantra that managed to override people's innate sense of caution about experimental products and lure them into accepting unproven injections.
- **Before our respondents or their children experienced vaccine injury themselves, the notion that vaccination might result in death or disability was nearly inconceivable.** As Karen put it, she "had no idea that anybody in the history of the world had been injured by a vaccine." Illustrating the spell cast by decades of indoctrination and media whiteouts, our respondents believed in the safety of vaccination even when they had encountered echoes of evidence to the contrary—a fellow first-grader's death following smallpox vaccination, a grandmother's child being "afflicted" after a "pox shot" during pregnancy, a friend's father developing GBS after a flu shot. Our stories also illustrate the powerful and tragic blunting of parental and personal instincts that allows vaccination to proceed even after "warning sign" reactions—red flags that doctors, too, routinely and unethically ignore.
- **In vaccination's "Russian roulette" context, not having a clear-cut adverse reaction in one set of circumstances tends to provide**

false reassurance that one will never experience an adverse reaction. Our COVID-vaccine-injured respondents all knew someone who had gone before them, seemingly uneventfully—in Maddie's and Mona's case, it was friends; in Suzanna's case, her husband; and in Junior's case, Ernesto himself. Parents of children injured by vaccines on the childhood vaccine schedule, too, had taken some of those same vaccines themselves (though not nearly as many). At the same time, our respondents' and their children's history of conditions like eczema, asthma, ADHD, headaches, and various and sundry medical-mystery afflictions—some of the 400 known adverse events listed in vaccine package inserts[509]—suggest that prior vaccination may not have been as benign as assumed, but information was lacking to connect the dots.

- As our nine stories show, **vaccine injury opens the door to a slew of additional pharmaceutical interventions.** This clever business model generates lucrative profits for industry—often creating "customers for life"—but rarely benefits the injured, who develop new problems that give rise to even more prescriptions and interventions. The rush to develop mRNA vaccines for problems created by the COVID injections illustrates this vicious cycle perfectly.
- **None of our respondents had ever considered the financial ramifications of a vaccine injury.** As some of our stories help illustrate, the financial impact often reverberates throughout the entire family, both immediate and extended, and not just over the short term but over the long term. In many instances, bankruptcy and foreshortened opportunities are very real possibilities. On the health front, the financial resources at an injured individual's disposal will dictate the extent to which they are able to explore non-pharmaceutical or holistic healing options. Those at the mercy of health insurance, whether public or private, will be limited to the tests and therapies that insurance is willing to cover. Those with savings or other resources convertible into cash will find those resources rapidly depleted.

- **Gaslighting of vaccine-injured individuals occurs on a massive scale and from all directions,** including by the medical profession, government regulators, the media, virtue-signaling members of the public posting on social media, and sometimes the injured party's own relatives. The general public's unwillingness to acknowledge the uncomfortable reality of vaccine injury leaves the injured socially isolated and is, as Suzanna emphasized, extremely hurtful—truly a case of "adding insult to injury."
- As another aspect of the medical gaslighting, our respondents emphasized that **many doctors cannot be trusted to chart their vaccine-injured patients' experiences accurately.** Over and over, individuals who accessed their own or their family member's medical records discovered that the charts distorted or omitted important facts about their health picture.
- As yet another facet of gaslighting, **the vaccine-injured will rarely if ever encounter medical professionals willing to admit to vaccination as a cause of their injury.** Instead, the era of mass vaccination has generated a plethora of medical diagnoses of "rare" diseases, filling diagnostic manuals full of "syndromes" and afflictions generally professed to be of "cause unknown." As Laura noted, most doctors will just "deal with the diagnoses in front of them" but will not risk their medical licenses—or, as Sheila put it, their bonuses—by overtly connecting the dots. During COVID, some observers have described "radical changes" in doctors' "attitudes toward data and each other," with the emergence of a previously "unthinkable" willingness to shame and report other doctors just for having differing opinions, abandoning their own "logic and curiosity."[510]
- Until the powerful films *Vaxxed: From Cover-up to Catastrophe* and *Vaxxed II: The People's Truth*,[511,512] **the vaccine injury phenomenon had attracted relatively few whistleblowers or even individuals willing to tell their story.** As Ernesto commented, he knows parents with COVID-vaccine-injured teens who are remaining

silent; FEMA's attempts to get him to change his son's death certificate made him wonder whether some families are being bought off.

SILENCE IS DEADLY

In 2006, discussing the earlier—and considerably suspect—epidemic known as AIDS,[513] veteran journalist Celia Farber described AIDS as "the story of the century," characterizing it as a "huge story that directly addresses truth in the information age, and how we know what we think we know." She also noted that AIDS raised questions about "the soul of science today in the pharmaceutical age and the biotech age." With the advent of the equally "huge" COVID story and the COVID-enabled rollout of dangerous new injections, questions about truth, official narratives, and the "soul of science" are even more urgent. At the height of AIDS, activists appropriated the slogan, "silence equals death," and it turns out that the same expression could be applied today to the topic of vaccination and its risks.

In early 2020, before the "pandemic" was on anyone's radar, retired physician Dr. Gary Kohls compiled over two dozen stories about vaccine injury and vaccine failure blatantly censored by the mainstream media, including stories about the medical profession's failure to acknowledge and report vaccine adverse events and industry's and regulators' "downplaying" of side effects that include infant death.[514] Kohls observed, "It is important for readers to understand that when true root causes of illnesses are denied or ignored by the medical profession . . . erroneous diagnoses will inevitably be made, treatments will be inevitably misguided and any chance for the prevention of future illnesses will be impossible."

In April 2022, Dr. Michael Yeadon, a highly qualified life sciences expert who worked for many years at Pfizer, went beyond condemning censorship to explain, in a piece titled "The Covid Lies," that "all the main narrative points about the coronavirus named SARS-CoV-2 are lies" and that "it is no longer possible to view the last two years as well-intentioned errors."[515] Supporting his statements scientifically with reference to peer-reviewed journal articles, Dr. Yeadon's inescapable conclusion is that the declared medical and public health emergency is merely a cover story for a wider agenda of totalitarian control over the population.

With the FDA now having green-lighted the COVID shots for our most precious and vulnerable—children under five and as young as six months of age—the time for silence, complacency, and self-interest is long past. An electrical engineer who submitted a public comment during the FDA's June 2022 deliberations about COVID jabs for six-month-olds told the FDA audience:

> "*[T]wo years ago I'd never heard of mRNA, but let me tell you what I've learned since. It starts with the shot you told us stays at the injection site. [...] You knew it didn't. [...] You knew the lipid nanocomplexes collect in the ovaries, where they have the potential to cause devastating effects on reproductive health, yet you did nothing. When women started complaining of menstrual problems you did nothing. [...] When people started dying of myocarditis you did nothing. [...] When the vaccinated repeatedly caught COVID and suffered reactivation of herpes, shingles, papillomavirus in unprecedented numbers you knew this was a massive problem, yet you did nothing. You knew that the mRNA stays around for months in the lymph node germinal centers causing T cell exhaustion. . . and then you ignored that massive safety signal. When you were warned about prion disease and amyloid. . . you did worse than nothing; you silenced those people who raise the alarms. You were informed of fraud in the vaccine studies, yet instead of investigating you colluded with the manufacturers to suppress trial data for 75 years.*"[516]

As Dr. Yeadon more straightforwardly concluded, "Serious crimes have obviously been committed." We must say "no" to coercion, mandates, concocted public health emergencies, and synthetic gene therapy injections. Family health and wealth—and the very future of mankind—are at stake.

Endnotes

[1] Greenfield B. Parents of vaccine-injured children speak out: "The guilt is huge." Yahoo! News, Feb. 17, 2015.
[2] Children's Health Defense. "Vaccine Mandates: An Erosion of Civil Rights?"—CHD's latest e-book available now." *The Defender*, Dec. 22, 2020.
[3] Lindsay R. Controlled opposition in the truth movement. Nature of Healing blog, May 9, 2017.
[4] Shilhavy B. The mainstream media has censored almost any negative information regarding vaccines. *The Christian Post*, Feb. 12, 2018.
[5] Farber C. "Bret Weinstein: 'It's a monster.'" *The Truth Barrier* [Substack], Jan. 31, 2022.
[6] Redshaw M. 17-year-old died "suddenly in sleep" 6 months after 2nd Pfizer shot. *The Defender*, Apr. 1, 2022.
[7] Helping people be seen, heard and believed after adverse vaccine reactions. Senator Ron Johnson, Adverse Vaccine Reactions Press Conference, July 2, 2021.
[8] Avendano M, Kawachi I. Why do Americans have shorter life expectancy and worse health than do people in other high-income countries? *Annu Rev Public Health*. 2014;35:307-325.
[9] Children's Health Defense. Dying young—falling life expectancy in the U.S. Oct. 1, 2019.
[10] Chronic conditions in America: price and prevalence. *RAND Review*, Jul. 12, 2017.
[11] Perrin JM, Anderson LE, Van Cleave J. The rise in chronic conditions among infants, children, and youth can be met with continued health system innovations. *Health Aff (Millwood)*. 2014;33(12):2099-2105.
[12] Bethell CD, Kogan MD, Strickland BB, et al. A national and state profile of leading health problems and health care quality for US children: key insurance disparities and across-state variations. *Acad Pediatr*. 2011;11(3 Suppl):S22-S33.
[13] Muennig PA, Reynolds M, Fink DS, et al. America's declining well-being, health, and life expectancy: not just a White problem. *Am J Public Health*. 2018;108(12):1626-1631.
[14] Samsel A, Seneff S. Glyphosate pathways to modern diseases VI: prions, amyloidoses and autoimmune neurological diseases. *J Biol Phys Chem*. 2017;17:8-32.
[15] Meldrum M. Opioids' long shadow. *AMA J Ethics*. 2020;22(1):E729-734.
[16] Ten thousand chemicals in food and food packaging: what are these substances doing to our children? Children's Health Defense, Aug. 23, 2018.

[17] Truesdale BC, Jencks C. The health effects of income inequality: averages and disparities. *Annu Rev Public Health*. 2016;37:413-430.

[18] Amadeo K. Medical bankruptcy and the economy. *The Balance*, Jan. 20, 2022.

[19] Saunders TJ, Vallance JK. Screen time and health indicators among children and youth: current evidence, limitations and future directions. *Appl Health Econ Health Policy*. 2017;15(3):323-331.

[20] *The Sickest Generation: The Facts Behind the Children's Health Crisis and Why it Needs to End* (second edition). Children's Health Defense, October 2020.

[21] CDC recommended vaccine schedule 1986 vs. 2019. Children's Health Defense, Sep. 5, 2019.

[22] https://canaryparty.org/commentary/the-canary-partys-new-viral-video-do-vaccines-cause-autism/

[23] Percentage of US children who have chronic health conditions on the rise. *Science Daily*, Apr. 30, 2016.

[24] Propp P, Becker A. Prevention of asthma: where are we in the 21st century? *Expert Rev Clin Immunol*. 2013;9(12):1267-1278.

[25] Jackson KD, Howie LD, Akinbami LJ. Trends in allergic conditions among children: United States, 1997–2011. *NCHS Data Brief*. 2013;121:1-8.

[26] Children's Health Defense. Vaccines can cause autoimmune diseases. Vaccine Secrets, Chapter 10. https://childrenshealthdefense.org/vaccine-secrets/video-chapters/

[27] Nieswand V, Richter M, Gossrau G. Epidemiology of headache in children and adolescents—another type of pandemia. *Curr Pain Headache Rep*. 2020;24(10):62.

[28] Children's Health Defense. Gaslighting autism families: CDC, media continue to obscure decades of vaccine-related harm. *The Defender*, Dec. 17, 2021.

[29] Auinger P, Lanphear BP, Kalkwarf HJ, Mansour ME. Trends in otitis media among children in the United States. *Pediatrics*. 2003;112:514-520.

[30] Data and statistics about ADHD. Centers for Disease Control and Prevention (last reviewed Sep. 23, 2021). https://www.cdc.gov/ncbddd/adhd/data.html

[31] Data and statistics on children's mental health. Centers for Disease Control and Prevention (last reviewed Mar. 4, 2022). https://www.cdc.gov/childrensmentalhealth/data.html

[32] Children's Health Defense. Is arthritis in childhood becoming the "new normal"—are vaccines to blame? Jan. 2, 2019.

[33] Skinner AC, Ravanbakht SN, Skelton JA, et al. Prevalence of obesity and severe obesity in US children, 1999–2016. *Pediatrics*. 2018;141(3):e20173459.

[34] Children's Health Defense. Vaxxed-Unvaxxed: The Science. https://childrenshealthdefense.org/wp-content/uploads/Vaxxed-Unvaxxed-Parts-I-XII.pdf

[35] Kennedy, Jr. RF. Vaccinated vs. unvaccinated—Part 11. Children's Health Defense, Dec. 7, 2020. https://childrenshealthdefense.org/news/vaccinated-vs-unvaccinated-part-11/

[36] Lyons-Weiler J, Thomas P. Relative incidence of office visits and cumulative rates of billed diagnoses along the axis of vaccination. *Int J Environ Res Public Health*. 2020;17(22):8674. [Retracted]

[37] Rikin S, Jia H, Vargas CY, et al. Assessment of temporally-related acute respiratory illness following influenza vaccination. *Vaccine*. 2018;36(15):1958-1964.

[38] Roosendaal study of vaccinated vs. unvaccinated children in the Netherlands: results survey. https://www.scribd.com/doc/60166949/Roosendaal-study-of-vaccinated-vs-unvaccinated-children-in-the-Netherlands-Results-Survey

[39] Mawson AR, Ray BD, Bhuiyan AR, Jacob B. Pilot comparative study on the health of vaccinated and unvaccinated 6- to 12-year-old U.S. children. *J Transl Sci.* 2017;3(3):1-12.

[40] Hurwitz EL, Morgenstern H. Effects of diphtheria-tetanus-pertussis or tetanus vaccination on allergies and allergy-related respiratory symptoms among children and adolescents in the United States. *J Manipulative Physiol Ther.* 2000;23(2):81-90.

[41] Enriquez R, Addington W, Davis F, et al. The relationship between vaccine refusal and self-report of atopic disease in children. *J Allergy Clin Immunol.* 2005;115(4):737-744.

[42] Hooker BS, Miller NZ. Analysis of health outcomes in vaccinated and unvaccinated children: developmental delays, asthma, ear infections and gastrointestinal disorders. *SAGE Open Med.* 2020;8:2050312120925344.

[43] Yamamoto-Hanada K, Pak K, Saito-Abe M, et al. Cumulative inactivated vaccine exposure and allergy development among children: a birth cohort from Japan. *Environ Health Prev Med.* 2020;25(1):27.

[44] McDonald KL, Huq SI, Lix LM, et al. Delay in diphtheria, pertussis, tetanus vaccination is associated with a reduced risk of childhood asthma. *J Allergy Clin Immunol.* 2008;121(3):626-631.

[45] Geier DA, Kern JK, Geier MR. A cross-sectional study of the relationship between reported human papillomavirus vaccine exposure and the incidence of reported asthma in the United States. *SAGE Open Med.* 2019;7:2050312118822650.

[46] Joshi AY, Iyer VN, Hertz MF, et al. Flu vaccination in asthmatics: does it work? *American Thoracic Society Conference Abstracts*, 2009.

[47] Shaheen SO, Aaby P, Hall AJ, et al. Measles and atopy in Guinea-Bissau. *Lancet.* 1996;347(9018):1792-1796.

[48] Geier DA, Kern JK, Homme KG, Geier MR. A cross-sectional study of the relationship between infant thimerosal-containing hepatitis B vaccine exposure and attention-deficit/hyperactivity disorder. *J Trace Elem Med Biol.* 2018;46:1-9.

[49] Geier DA, Hooker BS, Kern JK, et al. A two-phase study evaluating the relationship between thimersosal-containing vaccine administration and the risk for an autism spectrum disorder diagnosis in the United States. *Transl Neurodegener.* 2013;2(1):25.

[50] Gallagher CM, Goodman MS. Hepatitis B vaccination of male neonates and autism diagnosis, NHIS 1997-2002. *J Toxicol Environ Health A.* 2010;73(24):1665-1677.

[51] Generation zero: Thomas Verstraeten's first analyses of the link between vaccine mercury exposure and the risk of diagnosis of selected neuro-developmental disorders. Children's Health Defense, Jan. 10, 2018. https://childrenshealthdefense.org/government/foia/generation-zero-thomas-verstraetens-first-analyses-link-vaccine-mercury-exposure-risk-diagnosis-selected-neuro-developmental-disorders/

[52] DeStefano F, Bhasin TK, Thompson WW, et al. Age at first measles-mumps-rubella vaccination in children with autism and school-matched control subjects: a population-based study in metropolitan Atlanta. *Pediatrics.* 2004;113(2):259-266.

[53] Bardage C, Persson I, Ortqvist A, et al. Neurological and autoimmune disorders after vaccination against pandemic influenza A (H1N1) with a monovalent adjuvanted vaccine: population based cohort study in Stockholm, Sweden. *BMJ.* 2011;343:d5956.

[54] Pourcyrous M, Korones SB, Arheart KL, Bada HS. Primary immunization of premature infants with gestational age <35 weeks: cardiorespiratory complications and C-reactive protein responses associated with administration of single and multiple separate vaccines simultaneously. *J Pediatr*. 2007;151(2):167-172.

[55] Hviid A, Svanström H, Scheller NM, et al. Human papillomavirus vaccination of adult women and risk of autoimmune and neurological diseases. *J Intern Med*. 2018;283(2):154-165.

[56] Kharbanda EO, Vazquez-Benitez G, Lipkind HS, et al. Evaluation of the association of maternal pertussis vaccination with obstetric events and birth outcomes. *JAMA*. 2014;312(18):1897-1904.

[57] Thompson NP, Montgomery SM, Pounder RE, Wakefield AJ. Is measles vaccination a risk factor for inflammatory bowel disease? *Lancet*. 1995;345(8957):1071-1074.

[58] Pineton de Chambrun GP, Dauchet L, Gower-Rousseau C, et al. Vaccination and risk for developing inflammatory bowel disease: A meta-analysis of case-control and cohort studies. *Clin Gastroenterol Hepatol*. 2015;13(8):1405-1415.e1.

[59] Classen JB. Risk of vaccine induced diabetes in children with a family history of type 1 diabetes. *The Open Pediatric Medicine Journal*. 2008;2:7-10.

[60] Classen JB, Classen DC. Clustering of cases of type 1 diabetes mellitus occurring 2-4 years after vaccination is consistent with clustering after infections and progression to type 1 diabetes mellitus in autoantibody positive individuals. *J Pediatr Endocrinol Metab*. 2003;16(4):495-508.

[61] Classen DC, Classen JB. The timing of pediatric immunization and the risk of insulin-dependent diabetes mellitus. *Infect Dis Clin Pract (Baltim Md)*. 1997;6(7):449-454.

[62] Geier DA, Kern JK, Homme KG, Geier MR. Thimerosal exposure and disturbance of emotions specific to childhood and adolescence: a case-control study in the Vaccine Safety Datalink (VSD) database. *Brain Inj*. 2017;31(2):272-278.

[63] Goldman GS. Comparison of VAERS fetal-loss reports during three consecutive influenza seasons: was there a synergistic fetal toxicity associated with the two-vaccine 2009/2010 season? *Hum Exp Toxicol*. 2013;32(5):464-475.

[64] Kristensen I, Aaby P, Jensen H. Routine vaccinations and child survival: follow up study in Guinea-Bissau, West Africa. *BMJ*. 2000;321(7274):1435-1438.

[65] Miller NZ, Goldman GS. Infant mortality rates regressed against number of vaccine doses routinely given: is there a biochemical or synergistic toxicity? *Hum Exp Toxicol*. 2011;30(9):1420-1428.

[66] Aaby P, Jensen H, Gomes J, et al. The introduction of diphtheria-tetanus-pertussis vaccine and child mortality in rural Guinea-Bissau: an observational study. *Int J Epidemiol*. 2004;33(2):374-380.

[67] Mogensen SW, Andersen A, Rodrigues A, et al. The introduction of diphtheria-tetanus-pertussis and oral polio vaccine among young infants in an urban African community: a natural experiment. *EBioMedicine*. 2017;17:192-198.

[68] Aaby P, Ravn H, Roth A, et al. Early diphtheria-tetanus-pertussis vaccination associated with higher female mortality and no difference in male mortality in a cohort of low birthweight children: an observational study within a randomised trial. *Arch Dis Child*. 2012;97(8):685-691.

[69] Aaby P, Ravn H, Fisker AB, et al. Is diphtheria-tetanus-pertussis (DTP) associated with increased female mortality? A meta-analysis testing the hypotheses of sex-differential non-specific effects of DTP vaccine. *Trans R Soc Trop Med Hyg*. 2016;110(10):570-581.

[70] Moulton LH, Rahmathullah L, Halsey NA, et al. Evaluation of non-specific effects of infant immunizations on early infant mortality in a southern Indian population. *Trop Med Int Health.* 2005;10(10):947-955.

[71] Aaby P, Nielsen J, Benn CS, Trape JF. Sex-differential and non-specific effects of routine vaccinations in a rural area with low vaccination coverage: an observational study from Senegal. *Trans R Soc Trop Med Hyg.* 2015;109(1):77-84.

[72] Patel MM, López-Collada VR, Bulhöes MM, et al. Intussusception risk and health benefits of rotavirus vaccination in Mexico and Brazil. *N Engl J Med.* 2011;364(24):2283-2292.

[73] Yaju Y, Tsubaki H. Safety concerns with human papilloma virus immunization in Japan: analysis and evaluation of Nagoya City's surveillance data for adverse events. *Jpn J Nurs Sci.* 2019;16(4):433-449.

[74] Fisher MA, Eklund SA. Hepatitis B vaccine and liver problems in U.S. children less than 6 years old, 1993 and 1994. *Epidemiology.* 1999;10(3):337-339.

[75] Hernán MA, Jick SS, Olek MJ, Jick H. Recombinant hepatitis B vaccine and the risk of multiple sclerosis: a prospective study. *Neurology.* 2004;63(5):838-842.

[76] Miller E, Andrews N, Stellitano L, et al. Risk of narcolepsy in children and young people receiving AS03 adjuvanted pandemic A/H1N1 2009 influenza vaccine: retrospective analysis. *BMJ.* 2013;346:f794.

[77] Szakács A, Darin N, Hallböök T. Increased childhood incidence of narcolepsy in western Sweden after H1N1 influenza vaccination. *Neurology.* 2013;80(14):1315-1321.

[78] Mawson AR, Bhuiyan A, Jacob B, Ray BD. Preterm birth, vaccination and neurodevelopmental disorders: a cross-sectional study of 6- to 12-year-old vaccinated and unvaccinated children. *J Transl Sci.* 2017;3(3):1-8.

[79] Cowling BJ, Fang VJ, Nishiura H, et al. Increased risk of noninfluenza respiratory virus infections associated with receipt of inactivated influenza vaccine. *Clin Infect Dis.* 2012;54(12):1778-1783.

[80] Dierig A, Heron LG, Lambert SB, et al. Epidemiology of respiratory viral infections in children enrolled in a study of influenza vaccine effectiveness. *Influenza Other Respir Viruses.* 2014;8(3):293-301.

[81] Kelly H, Jacoby P, Dixon GA, et al. Vaccine effectiveness against laboratory-confirmed influenza in healthy young children: a case-control study. *Pediatr Infect Dis J.* 2011;30(2):107-111.

[82] Khurana S, Loving CL, Manischewitz J, et al. Vaccine-induced anti-HA2 antibodies promote virus fusion and enhance influenza virus respiratory disease. *Sci Transl Med.* 2013;5(200):200ra114.

[83] Geier DA, Young HA, Geier MR. Thimerosal exposure & increasing trends of premature puberty in the Vaccine Safety Datalink. *Indian J Med Res.* 2010;131:500-507.

[84] Geier DA, Kern JK, Geier MR. Premature puberty and thimerosal-containing hepatitis B vaccination: a case-control study in the Vaccine Safety Datalink. *Toxics.* 2018;6(4):67.

[85] Gallagher C, Goodman M. Hepatitis B triple series vaccine and developmental disability in US children aged 1–9 years. *Toxicological & Environmental Chemistry.* 2008;90(5):997-1008.

[86] Torch WC. Diphtheria-pertussis-tetanus (DPT) immunization: a potential cause of the sudden infant death syndrome (SIDS). *Neurology.* 1982;32(4, part 2):A169-170.

[87] Thompson WW, Price C, Goodson B, et al. Early thimerosal exposure and neuropsychological outcomes at 7 to 10 years. *N Engl J Med.* 2007;357(13):1281-1292.

[88] NCVIA: The legislation that changed everything—Conflicts of Interest Undermine Children's Health: Part II. Children's Health Defense, May 16, 2019.

[89] Richey W. Parents can't sue drug firms when vaccines cause harm, Supreme Court says. *The Christian Science Monitor*, Feb. 22, 2011.

[90] Fitts CA. The injection fraud—it's not a vaccine. The Solari Report, May 27, 2020. https://home.solari.com/deep-state-tactics-101-the-covid-injection-fraud-its-not-a-vaccine/

[91] Emergency use authorization. U.S. Food & Drug Administration, n.d. https://www.fda.gov/emergency-preparedness-and-response/mcm-legal-regulatory-and-policy-framework/emergency-use-authorization

[92] *The PREP Act and COVID-19: Limiting Liability for Medical Countermeasures.* Congressional Research Service, LSB10443, updated Jan. 13, 2022. https://crsreports.congress.gov/product/pdf/LSB/LSB10443/16

[93] Hals T. COVID-19 era highlights U.S. "black hole" compensation fund for pandemic vaccine injuries. Reuters, Aug. 21, 2020.

[94] Public Law 108-276. 108th Congress. July 21, 2004. https://www.congress.gov/108/plaws/publ276/PLAW-108publ276.pdf

[95] Iwry J. From 9/11 to COVID-19: a brief history of FDA emergency use authorization. COVID-19 and The Law (Harvard Law), Jan. 14, 2021.

[96] Iwry, 2021.

[97] ICAN vs. HHS: key legal win recasts vaccine debate. Informed Consent Action Network [press release], Sep. 14, 2018. https://www.prnewswire.com/news-releases/ican-vs-hhs-key-legal-win-recasts-vaccine-debate-300712629.html

[98] https://www.fda.gov/vaccines-blood-biologics/vaccine-adverse-events/vaccine-adverse-event-reporting-system-vaers-questions-and-answers

[99] Children's Health Defense. RFK, Jr. tells co-chair of new COVID advisory board: VAERS is broken, you can fix it. *The Defender*, Dec. 18, 2020.

[100] Eckert LO, Anderson BL, Gonik B, Schulkin J. Reporting vaccine complications: what do obstetricians and gynecologists know about the Vaccine Adverse Event Reporting System? *Infect Dis Obstet Gynecol.* 2013;2013:285257.

[101] https://vaers.hhs.gov/reportevent.html

[102] Lazarus R, Klompas M. *Electronic Support for Public Health—Vaccine Adverse Event Reporting System (ESP:VAERS).* Final report submitted by Harvard Pilgrim Health Care, Inc. to the Agency for Healthcare Research and Quality, 2011. https://digital.ahrq.gov/sites/default/files/docs/publication/r18hs017045-lazarus-final-report-2011.pdf

[103] Baker MA, Kaelber DC, Bar-Shain DS, et al. Advanced clinical decision support for vaccine adverse event detection and reporting. *Clin Infect Dis.* 2015;61(6):864-870.

[104] Children's Health Defense. RFK, Jr. tells co-chair of new COVID advisory board: VAERS is broken, you can fix it. *The Defender*, Dec. 18, 2020.

[105] https://www.hhs.gov/vaccines/national-vaccine-plan/goal-2/index.html

[106] Dr. Anthony Fauci: Risks from vaccines are "almost nonmeasurable." *Frontline*, Mar. 23, 2015.

[107] https://medalerts.org/vaersdb/findfield.php?TABLE=ON&GROUP1=AGE&EVENTS=ON&VAX_YEAR_HIGH=2020&VAX_MONTH_HIGH=11

[108] https://medalerts.org/vaersdb/findfield.php?TABLE=ON&GROUP1=AGE&EVENTS=ON&DIED=Yes&VAX_YEAR_HIGH=2020&VAX_MONTH_HIGH=11

[109] Vadalà M, Poddighe D, Laurino C, Palmieri B. Vaccination and autoimmune diseases: is prevention of adverse health effects on the horizon? *EPMA J.* 2017;8(3):295-311.
[110] Aquino MR, Bingemann TA, Nanda A, Maples KM. Delayed allergic skin reactions to vaccines. *Allergy Asthma Proc.* 2022;43(1):20-29.
[111] Kennedy Jr. RF. Eagle scout sues Merck, alleges Gardasil HPV vaccine destroyed his life. *The Defender*, Jan. 22, 2021.
[112] Classen JB, Classen DC. Public should be told that vaccines may have long term adverse effects. *BMJ.* 1999;318(7177):193.
[113] Package inserts & FDA product approvals. https://www.immunize.org/fda/
[114] Children's Health Defense. Read the fine print: vaccine package inserts reveal hundreds of medical conditions linked to vaccines. Apr. 14, 2020.
[115] https://www.fda.gov/media/74035/download
[116] Children's Health Defense. Read the fine print, part two—nearly 400 adverse reactions listed in vaccine package inserts. Aug. 14, 2020.
[117] https://www.bitchute.com/video/tQIFaGGsxjKV/
[118] Look WHO's talking! Vaccine scientists confirm major safety problems. Children's Health Defense, Jan. 16, 2020.
[119] Institute of Medicine. *The Childhood Immunization Schedule and Safety: Stakeholder Concerns, Scientific Evidence, and Future Studies.* Washington, DC: The National Academies Press; 2013.
[120] Redshaw M. FDA advisors unanimously endorse Pfizer, Moderna COVID shots for infants and young kids, ignore pleas to "first do no harm." *The Defender*, Jun. 15, 2022.
[121] U.S. Food and Drug Administration. Coronavirus (COVID-19) update: FDA limits use of Janssen COVID-19 vaccine to certain individuals. Press release, May 5, 2022.
[122] Redshaw M. Scientists warn Pfizer, Moderna vaccines may cause blood clots, too. *The Defender*, Apr. 16, 2021.
[123] Children's Health Defense. What you don't know could hurt you: Novavax's "loud-and-clear" nanoparticle adjuvant. *The Defender*, Jul. 20, 2022.
[124] Nova A, Kimball S. FDA authorizes emergency use for Novavax Covid-19 vaccine for ages 12 to 17. CNBC, Aug. 20, 2022.
[125] Mitteldorf J. Please don't call this "science": how FDA, CDC justified approval of Moderna's Spikevax. *The Defender*, Feb. 8, 2022.
[126] Middleton S. Pfizer, CDC lied to Americans, FDA-approved COVID shot exists on paper only. NaturalHealth365, Jun. 24, 2022.
[127] Nevradakis M. Exclusive: Whistleblowers accuse military of using Pfizer "Comirnaty" vaccine produced at facility not approved by FDA. *The Defender*, Aug. 10, 2022.
[128] COVID-19 vaccine candidates show gene therapy is a viable strategy. American Society of Gene + Cell Therapy, Nov. 17, 2020. https://asgct.org/research/news/november-2020/covid-19-moderna-nih-vaccine
[129] Opening ceremony World Health Summit 2021, speech Stefan Oelrich. Nov. 16, 2021. https://www.youtube.com/watch?v=IKBmVwuv0Qc
[130] Redshaw M. "We made a big mistake"—COVID vaccine spike protein travels from injection site, can cause organ damage. *The Defender*, Jun. 3, 2021.
[131] Redwood L. RFK, Jr. warned FDA three months ago about ingredient in Pfizer COVID vaccine that likely caused life-threatening reaction in two UK healthcare workers. *The Defender*, Dec. 11, 2020.

[132] Palmer M, Bhakdi S. Elementary, my dear Watson: why mRNA vaccines are a very bad idea. Doctors for Covid Ethics, Jan. 10, 2022.

[133] Palmer M, Bhakdi S, Wodarg W. Expert report on the Johnson & Johnson COVID-19 vaccine. Doctors for Covid Ethics, Feb. 9, 2022.

[134] Pfizer. *5.3.6 Cumulative Analysis of Post-Authorization Adverse Event Reports of PF-07302048 (BNT162B2) Received Through 28-Feb-2021.*

[135] Nevradakis M. Pfizer hired 600+ people to process vaccine injury reports, documents reveal. *The Defender*, Apr. 5, 2022.

[136] https://twitter.com/akheriaty/status/1510997946531135493

[137] CHD says Pfizer clinical trial data contradicts "safe and effective" government/industry mantra. Children's Health Defense [press release], Mar. 3, 2022.

[138] Nevradakis M. Pfizer classified almost all severe adverse events during COVID vaccine trials "not related to shots." *The Defender*, Jun. 21, 2022.

[139] Nevradakis M. FDA dumps more Pfizer documents: why were so many adverse events reported as "unrelated" to vaccine? *The Defender*, May 17, 2022.

[140] Excess mortality up 84% in millennials (ages 25-44) since 2021 vaccine rollout: former Blackrock executive. James Hill MD's Newsletter [Substack], Mar. 12, 2022.

[141] Children's Health Defense. Millennials besieged by chronic illness: from age 27, it's all "downhill." May 14, 2019.

[142] Thacker PD. Covid-19: researcher blows the whistle on data integrity issues in Pfizer's vaccine trial. *BMJ*. 2021;375:n2634.

[143] COVID vaccine trial whistleblower Brook Jackson files lawsuit against Pfizer for false claims. GreatGameIndia, Feb. 15, 2022.

[144] Kirsch S. Pfizer admits to COVID vaccine clinical trial fraud in federal court. Steve Kirsch's newsletter [Substack], Jun. 15, 2022.

[145] McLoone D. Doctor says Pfizer's COVID shot trial should be "null and void" after "twisting" data. *LifeSiteNews*, Jun. 24, 2022.

[146] Redshaw M. Nearly 30,000 deaths after COVID vaccines reported to VAERS, CDC data show. *The Defender*, May 13, 2022.

[147] https://www.cdc.gov/coronavirus/2019-ncov/vaccines/safety/vsafe.html

[148] Informed Consent Action Network. ICAN sues CDC to stop hiding v-safe data from the public. Legal Update, Dec. 29, 2021. https://www.icandecide.org/ican_press/ican-sues-cdc-to-stop-hiding-v-safe-data-from-the-public/

[149] Hause AM, Gee J, Baggs J, et al. COVID-19 vaccine safety in adolescents aged 12-17 years—United States, December 14, 2020–July 16, 2021. *MMWR Morb Mortal Wkly Rep*. 2021;70(31):1053-1058.

[150] Redshaw M. 21-year-old med student severely injured by Pfizer vaccine still waiting for response from government compensation program. *The Defender*, Feb. 25, 2022.

[151] Centers for Disease Control and Prevention. The Moderna COVID-19 vaccine's local reactions, systemic reactions, adverse events, and serious adverse events. https://www.cdc.gov/vaccines/covid-19/info-by-product/moderna/reactogenicity.html

[152] Shimabukuro TT, Kim SY, Myers TR, et al. Preliminary findings of mRNA Covid-19 vaccine safety in pregnant persons. *N Engl J Med*. 2021 Apr 21:NEJMoa2104983.

[153] Stroobandt S, Stroobandt R. Data of the COVID-19 mRNA-vaccine v-safe surveillance system and pregnancy registry reveals poor embryonic and second trimester fetal survival rate. Comment on Stuckelberger et al. SARS-CoV-2 vaccine willingness among

pregnant and breastfeeding women during the first pandemic wave: a cross-sectional study in Switzerland. *Viruses* 2021, *13*, 1199. *Viruses*. 2021;13(8):1545.

[154] Shilhavy B. 4,023 fetal deaths now recorded in VAERS following COVID-19 vaccines as US Appeals Court reinstates vaccine mandate for federal workers. *Health Impact News*, Apr. 9, 2022.

[155] Long P. COVID vaccines causing miscarriages, cancer and neurological disorders among military, DOD data show. *The Defender*, Jan. 26, 2022.

[156] Where is my period? With Mélodie Feron and Diane Protat. The Solari Report, Jun. 30, 2022. https://home.solari.com/coming-thursday-where-is-my-period-with-melodie-feron-and-diane-protat/

[157] Berenson A. URGENT: The Covid vaccine paper on declining sperm counts is even worse than it seems at first. Unreported Truths [Substack], Jun. 20, 2022.

[158] Kostoff RN, Calina D, Kanduc D, et al. Why are we vaccinating children against COVID-19? *Toxicol Rep*. 2021;8:1665-1684. [Retracted]

[159] What kind of shareholders hold the majority in RELX PLC's (LON:REL) shares? Yahoo!Finance, Jan. 17, 2022.

[160] Rohde W. How to file a petition in Vaccine Court. Children's Health Defense, Aug. 25, 2020.

[161] Rohde W. An unwelcome milestone: payouts for influenza vaccine injuries exceed $900 million. Children's Health Defense, Apr. 3, 2020.

[162] Flu shot injury compensation. Conway Homer PC, Attorneys at Law. https://ccandh.com/flu-shot-injury-compensation/

[163] Rohde, How to file a petition, 2020.

[164] Petitions filed, compensated and dismissed, by alleged vaccine, since the beginning of VICP, 10/01/1988 through 03/01/2022. Health Resources & Services Administration (HRSA). https://www.hrsa.gov/sites/default/files/hrsa/vaccine-compensation/data/vicp-stats-03-01-22.pdf

[165] *Vaccine Injury Compensation: Most Claims Took Multiple Years and Many Were Settled through Negotiation*. Government Accountability Office, GAO-15-142, Nov. 21, 2014. https://www.gao.gov/products/gao-15-142

[166] https://www.hrsa.gov/cicp/about

[167] Countermeasures Injury Compensation Program (CICP) Data: Aggregate Data as of March 1, 2022. Health Resources & Services Administration. https://www.hrsa.gov/cicp/cicp-data

[168] Nevradakis M. U.S. approves first injury claim for COVID countermeasure, as backlog grows to 4,000+ claims. *The Defender*, Dec. 22, 2021.

[169] Demasi M. Covid-19: is the US compensation scheme for vaccine injuries fit for purpose? *BMJ*. 2022;377:o919.

[170] Greene J. A "black hole" for COVID vaccine injury claims. Reuters, Jun. 29, 2021.

[171] Sigalos M. You can't sue Pfizer or Moderna if you have severe Covid vaccine side effects. The government likely won't compensate you for damages either. CNBC, Dec. 17, 2020.

[172] The true cost of autism. The Highwire, Episode 146, Jan. 16, 2020. https://thehighwire.com/videos/the-true-cost-of-autism-highwire-episode-146/

[173] Buescher AVS, Cidav Z, Knapp M, et al. Costs of autism spectrum disorders in the United Kingdom and the United States. *JAMA Pediatr*. 2014;168(8):721-728.

[174] Rogers T. "Autism tsunami": societal cost of autism in U.S. could hit $5.54 trillion by 2060, study predicts. *The Defender*, Aug. 11, 2021.

[175] Ganz ML. The lifetime distribution of the incremental societal costs of autism. *Arch Pediatr Adolesc Med.* 2007;161(4):343-349.

[176] Ozdemir NK, Koc M. Career adaptability of parents of children with autism spectrum disorder. *Curr Psychol.* 2022 Jan 27:1-14.

[177] McCall BP, Starr EM. Effects of autism spectrum disorder on parental employment in the United States: evidence from the National Health Interview Survey. *Community Work Fam.* 21(4):367-392.

[178] Callander EJ, Lindsay DB. The impact of childhood autism spectrum disorder on parent's labour force participation: can parents be expected to be able to re-join the labour force? *Autism.* 2018;22(5):542-548.

[179] The financial impact of an autism diagnosis. Autism Spectrum Disorder Foundation, n.d. http://myasdf.org/media-center/articles/the-financial-impact-of-an-autism-diagnosis/

[180] Ozdemir and Koc, 2022.

[181] Hartley SL, Barker ET, Seltzer MM, et al. The relative risk and timing of divorce in families of children with an autism spectrum disorder. *J Fam Psychol.* 2010;24(4):449-457.

[182] Does autism lead to more divorces? Claery & Hammond, LLP, Feb. 6, 2017. https://www.claerygreen.com/family-law-blog/2017/february/does-autism-lead-to-more-divorces-/

[183] University of Missouri-Columbia. Financial struggles plague families of children with autism. *ScienceDaily*, Feb. 29, 2008.

[184] "They don't want to see people like us." The Highwire, Episode 213, Apr. 30, 2021. https://thehighwire.com/videos/they-dont-want-to-see-people-like-us/

[185] Nevradakis M. Exclusive: Pilots injured by COVID vaccines speak out: "I will probably never fly again." *The Defender*, May 6, 2022.

[186] Spence WL. Workers' comp bill for vaccine injuries heads to House. Yahoo! News, Feb. 1, 2022.

[187] Smith A. Does workers' comp cover an employee's reaction to a COVID-19 vaccine? SHRM, Mar. 17, 2021.

[188] Workers could sue over vaccine mandates under Missouri bill. Associated Press, Apr. 21, 2022.

[189] Austin DA. Medical debt as a cause of consumer bankruptcy. *Maine Law Review.* 2015;67(1).

[190] Amadeo K. Medical bankruptcy and the economy. *The Balance*, Jan. 20, 2022.

[191] Fitts, The Injection Fraud, 2020.

[192] https://pandemic.solari.com/family-financial-disclosure-form-for-covid-19-injections/

[193] Children's Health Defense. COVID vaccine injuries—what's the financial risk? *The Defender*, Mar. 11, 2021.

[194] Children and families. National Alliance to End Homelessness, March 2021. https://endhomelessness.org/homelessness-in-america/who-experiences-homelessness/children-and-families/

[195] Bielenberg JE, Futrell M, Stover B, Hagopian A. Presence of any medical debt associated with two additional years of homelessness in a Seattle sample. *Inquiry.* 2020;57:0046958020923535.

[196] https://www.debt.org/medical/collections/
[197] Fox M. With rising inflation and a hot housing market, here's what you need to know about buying a home right now. CNBC, Nov. 23, 2021.
[198] Miami Rescue Mission: many families a paycheck away from being homeless. CBS Miami, Apr. 22, 2022.
[199] Redshaw M. Hundreds injured by COVID vaccines turn to GoFundMe for help with expenses. *The Defender*, Jul. 8, 2021.
[200] Houle JN, Keene DE. Getting sick and falling behind: health and the risk of mortgage default and home foreclosure. *J Epidemiol Community Health*. 2015;69(4):382-387.
[201] Sinnock B. First-mortgage default rate rises to a high last seen in early 2021. *National Mortgage News*, Feb. 16, 2022.
[202] Children's Health Defense. Diamond mine of data? Insurance companies report 40% increase in premature non-COVID deaths. *The Defender*, Jan. 5, 2022.
[203] Maready F. *The Autism Vaccine: The Story of Modern Medicine's Greatest Tragedy*. Wilmington, NC: Feels Like Fire; 2019.
[204] Blaxill MF. What's going on? The question of time trends in autism. *Public Health Rep*. 2004;119(6):536-551.
[205] Children's Health Defense. Gaslighting autism families: CDC, media continue to obscure decades of vaccine-related harm. *The Defender*, Dec. 17, 2021.
[206] Rossignol DA, Genuis SJ, Frye RE. Environmental toxicants and autism spectrum disorders: a systematic review. *Transl Psychiatry*. 2014;4(2):e360.
[207] Rappoport J. The big vaccine-autism lie. Jon Rappoport's Blog, May 20, 2018. https://blog.nomorefakenews.com/2018/05/20/the-big-vaccine-autism-lie/
[208] https://medalerts.org/vaersdb/findfield.php?TABLE=ON&GROUP1=AGE&EVENTS=ON&SYMPTOMS[]=Jaundice+%2810023126%29&SYMPTOMS[]=Jaundice+acholuric+%2810023128%29&SYMPTOMS[]=Jaundice+cholestatic-+%2810023129%29&SYMPTOMS[]=Jaundice+hepatocellular+%2810023136%29&SYMPTOMS[]=Jaundice+neonatal+%2810023138%29
[209] Celzo F, Buyse H, Welby S, Ibrahimi A. Safety evaluation of adverse events following vaccination with Havrix, Engerix-B or Twinrix during pregnancy. *Vaccine*. 2020;38(40):6215-6223.
[210] Garrido I, Lopes S, Simões MS, et al. Autoimmune hepatitis after COVID-19 vaccine—more than a coincidence. *J Autoimmun*. 2021;125:102741.
[211] Ghosh S. "Rare autoimmune hepatitis in people vaccinated with Covishield a concern." *The New Indian Express*, Jul. 11, 2021.
[212] Rela M, Jothimani D, Vij M, et al. Auto-immune hepatitis following COVID vaccination. *J Autoimmun*. 2021;123:102688.
[213] Mayer A. Groundbreaking study shows unvaccinated children are healthier than vaccinated children. *The Defender*, Dec. 7, 2020.
[214] Children's Health Defense. Acetaminophen—not worth the risk. Jul. 18, 2019.
[215] https://www.fda.gov/vaccines-blood-biologics/vaccines/pediarix
[216] Children's Health Defense. Read the fine print, part two—nearly 400 adverse reactions listed in vaccine package inserts. Aug. 14, 2020.
[217] Niyazov DM, Kahler SG, Frye RE. Primary mitochondrial disease and secondary mitochondrial dysfunction: importance of distinction for diagnosis and treatment. *Mol Syndromol*. 2016;7(3):122-137.

[218] Olmsted D. How autism happens: a conversation with Sheila Ealey. *Age of Autism*, Sep. 15, 2014.

[219] Warner A. Autism in the military. *Age of Autism*, Jul. 8, 2008. Comment by "Hera," posted Feb. 28, 2017.

[220] Ryan MAK, Smith TC, Sevick CJ, et al. Birth defects among infants born to women who received anthrax vaccine in pregnancy. *Am J Epidemiol.* 2008;168(4):434-442.

[221] Children's Health Defense. Historical lapses in public health ethics: will Gates-funded COVID vaccine human trials be business as usual? Jun. 18, 2020.

[222] Children's Health Defense. COVID vaccines don't make up for medical racism—they perpetuate it. *The Defender*, Mar. 15, 2021.

[223] Granpeesheh D, Tarbox J, Dixon DR. Applied behavior analytic interventions for children with autism: a description and review of treatment research. *Ann Clin Psychiatry.* 2009;21(3):162-173.

[224] https://www.stanleygreenspan.com/

[225] Autism Research Institute. *Treatment Options for Mercury/Metal Toxicity in Autism and Related Developmental Disabilities: Consensus Position Paper.* San Diego, CA: ARI, February 2005.

[226] Hammond JR. How a respected pediatrician lost his medical license—because he supported informed consent. *The Defender*, Jun. 14, 2021.

[227] Dr. Paul Thomas: CDC protects vaccine program, not kids' health. *The Defender*, Nov. 2, 2021.

[228] Kennedy Jr. RF. Join me in supporting Dr. Paul Thomas, a hero defending children's health. *The Defender*, Dec. 2, 2020.

[229] Is doctors' cash incentive sidelining the Hippocratic oath? Children's Health Defense, Sep. 19, 2019.

[230] Loehr J. Immunizations: how to protect patients and the bottom line. *Fam Pract Manag.* 2015;22(2):24-29.

[231] https://medicaid.ncdhhs.gov/providers/programs-and-services/long-term-care/community-alternatives-program-disabled-adults-capda

[232] https://www.navigatelifetexas.org/en/insurance-financial-help/texas-medicaid-waiver-programs-for-children-with-disabilities

[233] Conte L. Misconduct, mitochondria and the Omnibus Autism Proceedings. Children's Health Defense, Oct. 4, 2018.

[234] NCVIA: The legislation that changed everything—conflicts of interest undermine children's health: part II. Children's Health Defense, May 16, 2019.

[235] No authors listed. Rh-negative blood type in pregnancy. *J Midwifery Womens Health.* 2020;65(1):185-186.

[236] Pegoraro V, Urbinati D, Visser GHA, et al. Hemolytic disease of the fetus and newborn due to Rh(D) incompatibility: a preventable disease that still produces significant morbidity and mortality in children. *PloS One.* 2020;15(7):e0235807.

[237] Children's Health Defense. What polio vaccine injury looks like, decades later. Sep. 5, 2019.

[238] Hymolytic disease of the newborn (HDN). Stanford Children's Health. https://www.stanfordchildrens.org/en/topic/default?id=hemolytic-disease-of-the-newborn-90-P02368

[239] Kennedy Jr. RF. Mercury is not safe in any form: debunking the myths about thimerosal "safety." Children's Health Defense.

[240] Neighbor J. RhoGAM at 50: a Columbia drug still saving lives of newborns. Columbia University, Feb. 22, 2018.

[241] Children's Health Defense. Polio vaccine causing polio outbreaks in Africa, WHO admits. Sep. 3, 2020.

[242] Centers for Disease Control and Prevention. Recommended childhood immunization schedule—United States, 1995. *MMWR Recomm Rep*. 1995;44(No. RR-5):1-9.

[243] Testimony of Lyn Redwood, RN, MSN, President, Coalition for SafeMinds before the Subcommittee on Human Rights and Wellness Committee on Government Reform, U.S. House of Representatives, Sep. 8, 2004. Hearing: "Truth Revealed: New Scientific Discoveries Regarding Mercury in Medicine and Autism." https://www.safeminds.org/wp-content/uploads/2014/02/redwoodsafemindssept8testimonyfullfinal.pdf

[244] https://childrenshealthdefense.org/about-us/board-directors/

[245] Bernard S, Enayati A, Redwood L, et al. Autism: a novel form of mercury poisoning. *Med Hypotheses*. 2001;56(4):462-471.

[246] Guzzi G, Minoia C, Pigatto PD, Severi G. Methylmercury, amalgams, and children's health. *Environ Health Perspect*. 2006;114(3):A149.

[247] Children's Health Defense. The flawed logic of hepatitis B vaccine mandates. Jan. 31, 2019.

[248] Kennedy Jr. RF. Mercury is not safe in any form: debunking the myths about thimerosal "safety." Children's Health Defense.

[249] *Sykes v. Bayer*. Case No. 3:07CV660. https://www.dmlp.org/sites/citmedialaw.org/files/2008-01-22-Sykes%20v.%20Bayer%20Amended%20Complaint.pdf

[250] Kern JK, Geier DA, Deth RC, et al. Retracted article: Systematic assessment of research on autism spectrum disorder (ASD) and mercury reveals conflicts of interest and the need for transparency in autism research. *Sci Eng Ethics*. 2015;23(6):1689-1690.

[251] https://www.rxlist.com/rhogam-drug.htm

[252] Children's Health Defense. These "inactive" ingredients in COVID vaccines could trigger allergic reactions. *The Defender*, Mar. 12, 2021.

[253] The pharmaceutical industry's front men. Children's Health Defense, Jul. 23, 2019.

[254] González HFJ, Yengo-Kahn A, Englot DJ. Vagus nerve stimulation for the treatment of epilepsy. *Neurosurg Clin N Am*. 2019;30(2):219-230.

[255] What is the history of pertussis vaccine use in America? National Vaccine Information Center, n.d. https://www.nvic.org/vaccines-and-diseases/whooping-cough/vaccine-history.aspx

[256] Klein NP. Licensed pertussis vaccines in the United States. *Hum Vaccin Immunother*. 2014;10(9):2684-2690.

[257] Mold M, Umar D, King A, Exley C. Aluminium in brain tissue in autism. *J Trace Elem Med Biol*. 2018;46:76-82.

[258] Scientific review of Vaccine Safety Datalink Information. June 7-8, 2000. Norcross, GA: Simpsonwood Retreat Center. Page 105. https://childrenshealthdefense.org/wp-content/uploads/2016/10/The-Simpsonwood-Documents.pdf

[259] Children's Health Defense. CDC's vaccine "science"—a decades long trail of trickery. Jun. 25, 2019.

[260] "David Geier; There is No Safe Level Mercury." Evidence of Harm, Sep. 11, 2017. https://www.youtube.com/watch?v=LVGbIgFSWn4

[261] Dórea JG. Neurotoxic effects of combined exposures to aluminum and mercury in early life (infancy). *Environ Res*. 2020;188:109734.

[262] Greenspan SI, Wieder S. The Interdisciplinary Council on Developmental and Learning Disorders Diagnostic Manual for Infants and Young Children—an overview. *J Can Acad Child Adolesc Psychiatry*. 2008;17(2):76-89.

[263] https://www.ninds.nih.gov/health-information/disorders/apraxia

[264] https://www.psychologytoday.com/us/conditions/dyspraxia

[265] https://rarediseases.info.nih.gov/diseases/6855/landau-kleffner-syndrome/

[266] Children's Health Defense. One in nine adverse events reported after DTaP vaccination is serious—but CDC says, "don't worry, be happy." Jul. 31, 2018.

[267] Children's Health Defense. Amount of aluminum in infant vaccines "akin to a lottery," researchers say. *The Defender*, Apr. 21, 2021.

[268] Ghimire TR. The mechanisms of action of vaccines containing aluminum adjuvants: an *in vitro* vs *in vivo* paradigm. *Springerplus*. 2015;4:181.

[269] Mold M, Shardlow E, Exley C. Insight into the cellular fate and toxicity of aluminium adjuvants used in clinically approved human vaccinations. *Sci Rep*. 2016;6:31578.

[270] Mold M. Umar D, King A, Exley C. Aluminium in brain tissue in autism. *J Trace Elem Med Biol.* 2018;46:76-82.

[271] High aluminum found in autism brain tissue. Children's Health Defense, Nov. 28, 2017.

[272] Sawyer LA. Antibodies for the prevention and treatment of viral diseases. *Antiviral Res.* 2000;47(2):57-77.

[273] Families raise concern over mercury in vaccines. WRAL, Sep. 2, 2003.

[274] https://www.unlockingautism.org/about-us/

[275] https://nationalautismassociation.org/about-naa/honorary-board-members/

[276] https://safeminds.org/about/safeminds-activism/

[277] Teller M. Routinely risky: the shadow side of colonoscopy. Weston A. Price Foundation, Apr. 28, 2021.

[278] Type 1 diabetes on the rise in young children: is anyone paying attention? Children's Health Defense, Aug. 14, 2018.

[279] Childhood cancers, autism and environmental toxins. Children's Health Defense, Feb. 28, 2018.

[280] Eczema (atopic dermatitis) statistics. Allergy & Asthma Network, n.d. https://allergyasthmanetwork.org/what-is-eczema/eczema-statistics/

[281] Definition & facts for constipation in children. National Institute of Diabetes and Digestive and Kidney Diseases, n.d. https://www.niddk.nih.gov/health-information/digestive-diseases/constipation-children/definition-facts

[282] Brown S. Allergic reactions to a baby vaccination. Verywell Health, updated Apr. 20, 2022.

[283] https://www.cdc.gov/mmwr/preview/mmwrhtml/mm6105a5.htm#fig1

[284] Sears RW. *The Vaccine Book: Making the Right Decisions for Your Child.* Little, Brown and Company; 2011.

[285] Miller NZ. *Vaccines: Are They Really Safe and Effective?* New Atlantean Press; 2015.

[286] Miller NZ. *Miller's Review of Critical Vaccine Studies: 400 Important Scientific Papers Summarized for Parents and Researchers*. New Atlantean Press; 2016.

[287] https://www.mayoclinic.org/diseases-conditions/petit-mal-seizure/symptoms-causes/syc-20359683

[288] https://rarediseases.info.nih.gov/diseases/9684/sandifer-syndrome

[289] https://www.mayoclinic.org/diseases-conditions/pediatric-brain-tumor/symptoms-causes/syc-20361694

[290] https://medlineplus.gov/ency/article/007683.htm

[291] Merritt 2nd JL, Quinonez RA, Bonkowsky JL, et al. A framework for evaluation of the higher-risk infant after a brief resolved unexplained event. *Pediatrics.* 2019;144(2):e20184101.

[292] https://www.medalerts.org/vaersdb/findfield.php?TABLE=ON&GROUP1=AGE&E
VENTS=ON&SYMPTOMS[]=Apnoea+%2810002974%29&SYMPTOMS[]=Apn
oea+neonatal+%2810002976%29&SYMPTOMS[]=Apnoea+test+%2810053464%2-
9&SYMPTOMS[]=Apnoea+test+abnormal+%2810074913%29&SYMPTOMS[]=Ap
noeic+attack+%2810002977%29&SYMPTOMS[]=Breath+sounds+%2810064779%
29&SYMPTOMS[]=Breath+sounds+abnormal+%2810064780%29&SYMPTOMS[]
=Breath+sounds+absent+%2810062285%29&SYMPTOMS[]=Dyspnoea+%2810013
968%29&SYMPTOMS[]=Dyspnoea+at+rest+%2810013969%29&SYMPTOMS[]=
Dyspnoea+paroxysmal+nocturnal+%2810013974%29&VAX[]=6VAX-F&VAX[]=D
TAPHEPBIP&VAX[]=DTPHEP&VAX[]=DTPPVHBHPB&VAX[]=HBHEPB&VA
X[]=HEP&VAX[]=HEPAB&VAXTYPES=Hepatitis B&AGES[]=1&AGES[]=2

[293] https://www.medalerts.org/vaersdb/findfield.php?TABLE=ON&GROUP1=AGE&
EVENTS=ON&SYMPTOMS[]=Cyanosis+%2810011703%29&SYMPTOMS[]=
Cyanosis+central+%2810011704%29&SYMPTOMS[]=Cyanosis+neonatal+%281-
0011705%29&VAX[]=6VAX-F&VAX[]=DTAPHEPBIP&VAX[]=DTPHEP&VAX[
]=DTPPVHBHPB&VAX[]=HBHEPB&VAX[]=HEP&VAX[]=HEPAB&VAXTYPE
S=Hepatitis B&AGES[]=1&AGES[]=2

[294] Palmer A. Why has everyone seemingly forgotten how the immune system works? Children's Health Defense, Jul. 23, 2020.

[295] Children's Health Defense. Amount of aluminum in infant vaccines "akin to a lottery," researchers say. *The Defender*, Apr. 21, 2021.

[296] http://www.newyorkallianceforvaccinerights.org/

[297] Redshaw M. Supreme Court rejects appeal challenging New York's removal of religious exemption for schoolchildren. *The Defender*, May 26, 2022.

[298] Children's Health Defense. We've never seen vaccine injuries on this scale—why are regulatory agencies hiding COVID vaccine safety signals? *The Defender*, Aug. 12, 2021.

[299] Children's Health Defense. A tool of control: how health officials weaponize language to manage public perception of COVID vaccines. *The Defender*, Aug. 26, 2021.

[300] Redshaw M. Nearly 30,000 deaths after COVID vaccines reported to VAERS, CDC data show. *The Defender*, May 13, 2022.

[301] Mandel B. The risk to kids from COVID is miniscule. Do not let them mandate vaccines. *Newsweek*, Oct. 13, 2021.

[302] https://www.fda.gov/media/144416/download

[303] https://www.pfizer.com/science/coronavirus/vaccine/about-our-landmark-trial#

[304] Chow D. Pfizer launches trial to test Covid vaccine in children as young as 6 months. NBC, Mar. 25, 2021.

[305] https://www.fda.gov/media/148542/download

[306] https://www.fda.gov/media/153947/download

[307] https://www.fda.gov/media/154869/download

[308] Redshaw M. Reports of COVID vaccine injuries pass 1 million mark, FDA signs off on Pfizer booster for kids 12 and up. *The Defender*, Jan. 3, 2022.

[309] https://www.fda.gov/media/157364/download
[310] Redshaw M. FDA authorizes Pfizer booster for kids 5 to 11, bypasses advisory panel. *The Defender*, May 17, 2022.
[311] https://www.fda.gov/news-events/press-announcements/coronavirus-covid-19-update-fda-authorizes-moderna-and-pfizer-biontech-covid-19-vaccines-children
[312] Expert statement regarding the use of Moderna COVID-19-mRNA-vaccine in children. Doctors for Covid Ethics, Apr. 10, 2021. https://doctors4covidethics.org/expert-statement-regarding-the-use-of-moderna-covid-19-mrna-vaccine-in-children/
[313] Palmer M, Bhakdi S, Hockertz S. Expert evidence regarding Comirnaty (Pfizer) COVID-19 mRNA vaccine for children. Doctors for Covid Ethics, Mar. 7, 2021. https://doctors4covidethics.org/expert-evidence-regarding-comirnaty-covid-19-mrna-vaccine-for-children/
[314] Notice of liability for harm and death to children served on all members of the European Parliament. Doctors for Covid Ethics, May 19, 2021. https://doctors4covidethics.org/notice-of-liability-for-harm-and-death-to-children-served-on-all-members-of-the-european-parliament/
[315] https://www.realnotrare.com/post/maddie-de-garay
[316] Redshaw M. U.S. Sen. Johnson holds news conference with families injured by COVID vaccines, ignored by medical community. *The Defender*, Jun. 29, 2021.
[317] Redshaw M. Vaccine-injured speak out, feel abandoned by government who told them COVID shot was safe. *The Defender*, Nov. 3, 2021.
[318] Rigged: Maddie's Story. The Highwire, Episode 280, Aug. 13, 2022. https://thehighwire.com/videos/rigged-maddies-story/
[319] Research annual report: a message from Mark Jahnke, Nancy Krieger Eddy, PhD, and Michael Fisher. Cincinnati Children's, n.d. https://www.cincinnatichildrens.org/research/cincinnati/annual-report/2020/board
[320] https://www.drugs.com/vyvanse.html
[321] Silver L. How to tame the tics associated with ADHD medication. *ADDitude*, Apr. 6, 2022.
[322] New research suggests link between vaccine ingredients and autism, ADHD. *Newswise*, Feb. 3, 2004.
[323] Geier DA, Kern JK, Homme KG, Geier MR. Abnormal brain connectivity spectrum disorders following thimerosal administration: a prospective longitudinal case-control assessment of medical records in the Vaccine Safety Datalink. *Dose Response.* 2017;15(1):1559325817690849.
[324] ADHD: alarms raised; risks ignored. Children's Health Defense, Nov. 19, 2019.
[325] Mawson AR, Ray BD, Bhuiyan AR, Jacob B. Pilot comparative study on the health of vaccinated and unvaccinated 6- to 12-year old U.S. children. *J Transl Sci.* 2017;3.
[326] Leslie DL, Kobre RA, Richmand BJ, et al. Temporal association of certain neuropsychiatric disorders following vaccination of children and adolescents: a pilot case-control study. *Front Psychiatry*. 2017;8:3.
[327] Read the fine print, part two—nearly 400 adverse reactions listed in vaccine package inserts. Children's Health Defense, Aug. 14, 2020.
[328] https://www.fda.gov/media/75718/download
[329] https://www.fda.gov/media/75195/download
[330] Yamamoto-Hanada et al., 2020.
[331] https://www.fda.gov/media/90064/download

[332] Read the fine print, part two—nearly 400 adverse reactions listed in vaccine package inserts. Children's Health Defense, Aug. 14, 2020.

[333] Cherney K. What is dermatographia? *Healthline*, Apr. 18, 2019.

[334] Lloreda CL. Covid-19 kept families and caregivers out of hospitals. Some doctors think that shouldn't happen again. *Stat*, Jul. 20, 2021.

[335] Jekielek J. The vaccine-injured and their fight for treatment, transparency—trial participants Maddie and Stephanie de Garay and Brianne Dressen. American Thought Leaders [Epoch TV], Jan. 27, 2022.

[336] Mom: Pfizer, FDA, CDC concealed adverse reactions ahead of vaccine clinical trials; have yet to respond. *WorldTribune*, May 8, 2022.

[337] https://www.realnotrare.com/post/maddie-de-garay

[338] Bandcroft J. Do you know about Maddie de Garay? Plant Powered Radio Blog [Substack], Jan. 20, 2022.

[339] Jekielek, 2022.

[340] Butler M, Coebergh J, Safavi F, et al. Functional neurological disorder after SARS-CoV-2 vaccines: two case reports and discussion of potential public health implications. *J Neuropsychiatry Clin Neurosci.* 2021;33(4):345-348.

[341] Ahead of the curve: The vaccine injured unite. The Highwire, Episode 226, Aug. 2, 2021. https://thehighwire.com/videos/the-vaccine-injured-unite/

[342] Begue I, Adams C, Stone J, Perez DL. Structural alterations in functional neurological disorder and related conditions: a software *and* hardware problem? *Neuroimage Clin.* 2019;22:101798.

[343] https://rarediseases.org/rare-diseases/fnd/

[344] George J. Vaccine effect of functional neurological disorder? *MedPage Today*, Aug. 20, 2021.

[345] https://www.cdc.gov/mis/mis-c.html

[346] Therapeutic management of hospitalized pediatric patients with multisystem inflammatory syndrome in children (MIS-C) (with discussion on multisystem inflammatory syndrome in adults [MIS-A]). COVID-19 Treatment Guidelines, National Institutes of Health, updated Feb. 24, 2022.

[347] Redshaw M. 8-year-old boy dies of MIS 7 days after Pfizer vaccine, VAERS report shows. *The Defender*, Feb. 25, 2022.

[348] https://www.cincinnatichildrens.org/bio/f/robert-frenck

[349] Frenck RW Jr., Klein NP, Kitchin N, et al. Safety, immunogenicity, and efficacy of the BNT162b2 Covid-19 vaccine in adolescents. *N Engl J Med.* 2021;385:239-250.

[350] Cincinnati Children's Frenck lead author of COVID-19 vaccine study in NEJM. Cincinnati Children's Hospital Medical Center, May 27, 2021.

[351] https://rarediseases.org/rare-diseases/fnd/

[352] https://www.drugs.com/gabapentin.html

[353] https://www.drugs.com/lyrica.html

[354] Jekielek, 2022.

[355] https://www.lifefunder.com/maddie

[356] Brianne Dressen severe ongoing adverse reactions from AstraZeneca clinical trial of 2020. CovidVaccineInjuries.com, Mar. 16, 2022. https://community.covidvaccineinjuries.com/brianne-dressen-severe-ongoing-adverse-reactions-from-astrazeneca-clinical-trial-of-2020/

[357] https://react19.org/about-react-19/

[358] https://react19.org/donate/

[359] https://motherhoodcommunity.com/elecare-recall/

[360] Jekielek, 2022.

[361] "The People's Testaments" Episode 27: Ernest Ramirez—the full story. CHD. TV, Mar. 21, 2022. https://live.childrenshealthdefense.org/shows/the-peoples-testaments/9qvGaO2sMl

[362] Grieving father Ernest Ramirez shares heartbreaking story of his teen son's death 5 days after Pfizer vaccine. Circle of Mamas, Nov. 3, 2021.

[363] Hundreds of professional athletes collapsing on field, dying from mysterious heart complications. One America News Network, Apr. 10, 2022.

[364] Conradson J. Update: A jaw-dropping 769 athletes have collapsed while competing over the past year—"avg. age of players suffering cardiac arrest is just 23"—(video). *Gateway Pundit*, Apr. 8, 2022.

[365] Let's Go to the Movies: Week of May 2, 2022: On the record—an interview with Matt Le Tissier. The Solari Report, Apr. 30, 2022. https://home.solari.com/lets-go-to-the-movies-week-of-may-2-2022-on-the-record-an-interview-with-matt-le-tissier/

[366] Fogoros RN. Myocarditis exercise recommendations and the importance of following activity restrictions. Verywell Health, May 11, 2022.

[367] https://twitter.com/IvoryHecker/status/1431284807049359364

[368] Gore L. TV reporter Ivory Hecker, fired after accusing station of muzzling her, releases recordings. Advance Local, Jun. 17, 2021.

[369] McBride J. Ivory Hecker: 5 fast facts you need to know. Heavy.com, Jun. 16, 2021.

[370] De Anda M. Moderna vaccine trials for children to start soon in Hidalgo County. KRGV.com, Sep. 26, 2021.

[371] Cole A. DHR Health continues vaccinating against COVID-19, 292,300 shots administered. KRGV.com, Mar. 9, 2022.

[372] Garcia C. Annual Covid vaccines possible, Hidalgo County health authority says. KRGV.com, Apr. 14, 2022.

[373] De Anda M. Hidalgo County officials urge parents to vaccinate their children. KRGV.com, Jul. 20, 2021.

[374] Hajjo R, Sabbah DA, Bardaweel SK, Tropsha A. Shedding the light on post-vaccine myocarditis and pericarditis in COVID-19 and non-COVID-19 vaccine recipients. *Vaccines (Basel)*. 2021;9(10):1186.

[375] Mei R, Raschi E, Forcesi E, et al. Myocarditis and pericarditis after immunization: gaining insights through the Vaccine Adverse Event Reporting System. *Int J Cardiol.* 2018;273:183-186.

[376] Kumar V, Sidhu N, Roy S, Gaurav K. Myocarditis following diphtheria, whole-cell pertussis, and tetanus toxoid vaccination in a young infant. *Ann Pediatr Cardiol.* 2018;11(2):224-226.

[377] Ho JSY, Sia CH, Ngiam JN, et al. A review of COVID-19 vaccination and the reported cardiac manifestations. *Singapore Med J.* 2021 Nov 19.

[378] Children's Health Defense. Vaccine-induced myocarditis injuring record number of young people. Will shots also bankrupt families? *The Defender*, Jan. 31, 2022.

[379] Bhakdi S, Burkhardt A. On COVID vaccines: why they cannot work, and irrefutable evidence of their causative role in deaths after vaccination. Doctors for Covid Ethics, Dec. 15, 2021. https://doctors4covidethics.org/on-covid-vaccines-why-they-cannot-work-and-irrefutable-evidence-of-their-causative-role-in-deaths-after-vaccination/

[380] Ho et al., 2021.
[381] Shilhavy B. Government VAERS data reveal 15,600% increase in heart disease among under 30-year-olds following COVID-19 vaccination. *Health Impact News*, Jan. 28, 2022.
[382] Vidula MK, Ambrose M, Glassberg H, et al. Myocarditis and other cardiovascular complications of the mRNA-based COVID-19 vaccines. *Cureus*. 2021;13(6):e15576.
[383] Redshaw M. Vaccine-injured speak out, feel abandoned by government who told them COVID shot was safe. *The Defender*, Nov. 3, 2021.
[384] https://campaigns.dailyclout.io/invitation?type=brand&code=&brand=cc3b3e5a-6536-4738-8ed6-5ee368c67240
[385] Flowers C. Pfizer vaccine: FDA fails to mention risk of heart damage in teens. Apr. 7, 2022. *Daily Clout*, Apr. 7, 2022.
[386] Wolf N. Dear friends, sorry to announce a genocide. Outspoken with Dr Naomi Wolf [Substack], May 29, 2022.
[387] https://www.fda.gov/media/148542/download
[388] Redshaw M. Young boy died of myocarditis after Pfizer vaccine, says CDC before signing off on 3rd shot for kids 5-11. *The Defender*, May 27, 2022.
[389] "The People's Testaments" episode 27: Ernest Ramirez—the full story. CHD. TV, Mar. 21, 2022. https://live.childrenshealthdefense.org/shows/the-peoples-testaments/9qvGaO2sMl
[390] https://www.fema.gov/disaster/coronavirus/economic/funeral-assistance
[391] Jackson T. Update: GoFundMe de-platforms dad whose only son died after Pfizer jab. *LifeSiteNews*, Sep. 15, 2021.
[392] Miltimore J. Why GoFundMe deleted this grieving father's fundraiser after his son's death. Foundation for Economic Education (FEE), Sep. 24, 2021.
[393] https://www.lifefunder.com/pfizervictim
[394] https://www.givesendgo.com/G28FK
[395] Tikkanen R, Abrams MK. *U.S. Health Care from a Global Perspective, 2019: Higher Spending, Worse Outcomes?* The Commonwealth Fund, Jan. 30, 2020.
[396] Raghupathi W, Raghupathi V. An empirical study of chronic diseases in the United States: a visual analytics approach to public health. *Int J Environ Res Public Health*. 2018;15(3):431.
[397] Chronic conditions in America: price and prevalence. *RAND Review*, Jul. 12, 2017.
[398] Raghupathi and Raghupathi, 2018.
[399] Percent of U.S. adults 55 and over with chronic conditions. CDC, National Center for Health Statistics (page last reviewed Nov. 6, 2015). https://www.cdc.gov/nchs/health_policy/adult_chronic_conditions.htm
[400] Tikkanen and Abrams, 2020.
[401] The CDC's influenza math doesn't add up: exaggerating the death toll to sell flu shots. Children's Health Defense, Oct. 9, 2018.
[402] Children's Health Defense. Let's learn from South Korea's aggressive rollout of its flu vaccine campaign. *The Defender*, Oct. 29, 2020.
[403] Rohde W. An unwelcome milestone: payouts for influenza vaccine injuries exceed $900 million. Children's Health Defense, Apr. 3, 2020.
[404] Doshi PN. *Influenza: A Study of Contemporary Medical Politics*. Massachusetts Institute of Technology; 2011.

[405] Committee to Review Adverse Effects of Vaccines; Institute of Medicine; Stratton K, Ford A, Rusch E, et al., editors. "Influenza Vaccine." Chapter 6 in *Adverse Effects of Vaccines: Evidence and Causality.* Washington, DC: National Academies Press (US); Aug. 25, 2011.

[406] Lanza GA, Barone L, Scalone G, et al. Inflammation-related effects of adjuvant influenza A vaccination on platelet activation and cardiac autonomic function. *J Intern Med.* 2011;269(1):118-125.

[407] Kennedy Jr. RF. CDC study shows up to 7.7-fold greater odds of miscarriage after influenza vaccine. Children's Health Defense, Sep. 19, 2017.

[408] Christian LM, Iams JD, Porter K, Glaser R. Inflammatory responses to trivalent influenza virus vaccine among pregnant women. *Vaccine*. 2011;29(48):8982-8987.

[409] Unwin C, Blatchley N, Coker W, et al. Health of UK servicemen who served in Persian Gulf War. *Lancet.* 1999;353(9148):169-178.

[410] Hotopf M, David A, Hull L, et al. Role of vaccinations as risk factors for ill health in veterans of the Gulf war: cross sectional study. *BMJ.* 2000;320(7246):1363-1367.

[411] Steele L. Prevalence and patterns of Gulf War illness in Kansas veterans: Association of symptoms with characteristics of person, place, and time of military service. *Am J Epidemiol.* 2000;152(10):992-1002.

[412] *Anthrax Vaccine: GAO's Survey of Guard and Reserve Pilots and Aircrew.* U.S. Government Accountability Office, GAO-02-445, Sep. 20, 2002. https://www.gao.gov/products/gao-02-445

[413] Long P. What COVID vaccine policymakers can learn from botched military anthrax vaccine program. *The Defender*, Oct. 21, 2020.

[414] Children's Health Defense. Vaccines are sabotaging the immune system. Shingles may hold some answers. *The Defender*, Dec. 9, 2021.

[415] https://www.cdc.gov/vaccines/parents/travel-vaccines.html

[416] The yellow fever vaccine: more questions than answers. Children's Health Defense, Feb. 21, 2019.

[417] De Menezes Martins R, da Luz Fernandes Leal M, Homma A. Serious adverse events associated with yellow fever vaccine. *Hum Vaccin Immunother*. 2015;11(9):2183-2187.

[418] https://www.lisaeden.com/

[419] Yamamoto-Hanada et al., 2020.

[420] Cocores A, Monteith T. Post-vaccination headache reporting trends according to the Vaccine Adverse Events Reporting System (VAERS) (P1.147). *Neurology.* 2016;86(16 Suppl).

[421] Krogsgaard LW, Helmuth IG, Bech BH, et al. Are unexplained adverse health events following HPV vaccination associated with infectious mononucleosis? – A Danish nationwide matched case-control study. *Vaccine*. 2020;38(35):5678-5684.

[422] Committee on Infectious Diseases and Committee on Environmental Health. Thimerosal in vaccines—an interim report to clinicians. *Pediatrics*. 1999;104(3):570-574.

[423] Zafrir Y, Agmon-Levin N, Paz Z, et al. Autoimmunity following hepatitis B vaccine as part of the spectrum of "Autoimmune (Auto-inflammatory) Syndrome induced by Adjuvants" (ASIA): analysis of 93 cases. *Lupus*. 2012;21(2):146-152.

[424] Agmon-Levin N, Zafrir Y, Kivity S, et al. Chronic fatigue syndrome and fibromyalgia following immunization with the hepatitis B vaccine: another angle of the "autoimmune

(auto-inflammatory) syndrome induced by adjuvants" (ASIA). *Immunol Res*. 2014;60(2-3):376-383.
[425] Watad A, Sharif K, Shoenfeld Y. The ASIA syndrome: basic concepts. *Mediterr J Rheumatol.* 2017;28(2):64-69.
[426] https://www.rxlist.com/orimune-drug.htm
[427] Dimova S, Hoet PHM, Dinsdale D, Nemery B. Acetaminophen decreases intracellular glutathione levels and modulates cytokine production in human alveolar macrophages and type II pneumocytes in vitro. *Int J Biochem Cell Biol.* 2005;37(8):1727-1737.
[428] Why Tylenol should be avoided by everyone. Dr. G, Mar. 6, 2016. https://drsteveng.com/why-tylenol-should-be-avoided-by-everyone/
[429] Multiple chemical sensitivity treatments in chronic fatigue syndrome. Quit Chronic Fatigue, Mar. 14, 2019.
[430] International Academy of Oral Medicine & Toxicology. https://iaomt.org/
[431] Haelle T. California vaccination bill SB 277 signed by governor, becomes law. *Forbes*, Jun. 30, 2015.
[432] https://physiciansforinformedconsent.org/
[433] Families sue New York State to stop the repeal of the religious exemption. Children's Health Defense, Jul. 12, 2019.
[434] Trust the Moms - Lisa Eden on Vaccine Injury. Apr. 30, 2019. https://www.youtube.com/watch?v=iBWZimk6QLg&ab_channel=SallieO.Elkordy
[435] Roos R. Study: flu shots in elderly don't cut mortality rate. Center for Infectious Disease Research and Policy (CIDRAP), Feb. 16, 2005.
[436] Anderson ML, Dobkin C, Gorry D. The effect of influenza vaccination for the elderly on hospitalization and mortality: an observational study with a regression discontinuity design. *Ann Intern Med.* 2020;172(7):445-452.
[437] Menge M. BREAKING: Fifth largest life insurance company in the US paid out 163% more for deaths of working people ages 18-64 in 2021 – total claims/benefits up $6 BILLION. Crossroads Report [Substack], Jun. 15, 2022.
[438] Children's Health Defense. Diamond mine of data? Insurance companies report 40% increase in premature non-COVID deaths. *The Defender*, Jan. 5, 2022.
[439] Hancock S. 40% rise nationwide in excess deaths among 18- to 49-year-olds, CDC data show. *The Defender*, Jan. 20, 2022.
[440] Bruce A. Edward Dowd on future recession, shocking findings in the CDC Covid data and democide. LewRockwell.com, Mar. 12, 2022.
[441] Alexander P. US military doctor testifies she was ordered by govn administration to "cover up" vaccine injuries. Alexander COVID News evidence-based medicine convoy mandate [Substack], Apr. 3, 2022.
[442] https://fred.stlouisfed.org/series/LNU00074597
[443] Farber C. Mayday! Dowd: Three million extra Americans on disability since vaccine rollout. The Truth Barrier [Substack], Jun. 13, 2021.
[444] https://www.realnotrare.com/post/mona-hasegawa
[445] Mangiaracina E. Woman ends up in wheelchair with neurological damage after submitting to COVID shot. *LifeSiteNews*, Nov. 5, 2021.
[446] https://my.clevelandclinic.org/health/diseases/16872-chronic-venous-insufficiency-cvi
[447] https://www.mayoclinic.org/diseases-conditions/transient-ischemic-attack/symptoms-causes/syc-20355679
[448] https://www.britannica.com/topic/Hippocratic-oath

[449] https://my.clevelandclinic.org/health/diseases/16560-postural-orthostatic-tachycardia-syndrome-pots
[450] University of Toledo. New research adds to evidence that POTS may be an autoimmune disorder. *Medical Xpress*, Mar. 16, 2021.
[451] Kennedy Jr. RF. Fourth Gardasil lawsuit against Merck alleges its HPV vaccine caused debilitating injuries. *The Defender*, Nov. 18, 2020.
[452] Tomljenovic L, Colafrancesco S, Perricone C, Shoenfeld Y. Postural orthostatic tachycardia with chronic fatigue after HPV vaccination as part of the "Autoimmune/Auto-inflammatory Syndrome Induced by Adjuvants": case report and literature review. *J Investig Med High Impact Case Rep*. 2014;2(1):2324709614527812.
[453] Ward D, Thorsen NM, Frisch M, et al. A cluster analysis of serious adverse event reports after human papillomavirus (HPV) vaccination in Danish girls and young women, September 2009 to August 2017. *Eurosurveillance*. 2019;24(19):pii=1800380.
[454] https://www.mayoclinic.org/documents/honor-roll-pdf/DOC-20079530
[455] Butts BN, Fischer PR, Mack KJ. Human papillomavirus vaccine and postural orthostatic tachycardia syndrome: a review of current literature. *J Child Neurol*. 2017;32(11):956-965.
[456] Reddy S, Reddy S, Arora M. A case of postural orthostatic tachycardia syndrome secondary to the messenger RNA COVID-19 vaccine. *Cureus*. 2021;13(5):e14837.
[457] Dahan S, Tomljenovic L, Shoenfeld Y. Postural orthostatic tachycardia syndrome (POTS) – a novel member of the autoimmune family. *Lupus*. 2016;25(4):339-342.
[458] Nevradakis M. Exclusive: 29-year-old's career came "crashing" down after Pfizer COVID vaccine injury. *The Defender*, Jun. 16, 2022.
[459] Reddy et al., 2021.
[460] https://react19.wpengine.com/wp-content/uploads/2022/04/022-VaccineAEs-for-practitioners-2-3-1.pdf
[461] Schofield JR, Chemali KR. How we treat autoimmune small fiber neuropathy with immunoglobulin therapy. *Eur Neurol*. 2018;80:304-310.
[462] Fernando J. Corporate social responsibility (CSR). Investopedia, Mar. 7, 2022.
[463] https://www.mayoclinic.org/diseases-conditions/sjogrens-syndrome/symptoms-causes/syc-20353216
[464] What is small fiber neuropathy? Ameripharma Specialty Care, Sep. 5, 2021.
[465] Hogan L. What is small fiber sensory neuropathy? WebMD, Nov. 9, 2021.
[466] Devigili G, Rinaldo S, Lombardi R, et al. Diagnostic criteria for small fibre neuropathy in clinical practice and research. *Brain*. 2019;142(12):3728-3736.
[467] https://www.hopkinsmedicine.org/health/conditions-and-diseases/polymyositis
[468] Nienaber P, Staver M, Cox G. This story is just "too threatening" for our government. Faith & Liberty, Jul. 16, 2021.
[469] Marks D. Media censoring vaccine injury stories, Brianne Dressen tells RFK, Jr. *The Defender*, Jan. 27, 2022.
[470] https://www.guptaprogram.com/
[471] https://www.reiki.org/faqs/what-reiki
[472] https://www.myvalleycryo.com/red-light-therapy/
[473] Laughing all the way to the bank: vaccine makers and liability protection—Conflicts of Interest Undermine Children's Health: Part III. Children's Health Defense, May 23, 2019.